INTERNATIONAL SERIES OF MONOGRAPHS IN
ANALYTICAL CHEMISTRY

GENERAL EDITORS: R. BELCHER AND H. FREISER

VOLUME 50

N-BENZOYLPHENYLHYDROXYLAMINE AND ITS ANALOGUES

N-BENZOYLPHENYLHYDROXYLAMINE AND ITS ANALOGUES

BY

A. K. MAJUMDAR

Jadavpur University, Calcutta-32, India

PERGAMON PRESS

Oxford · New York · Toronto
Sydney · Braunschweig

Pergamon Press Ltd., Headington Hill Hall, Oxford
Pergamon Press Inc., Maxwell House, Fairview Park, Elmsford,
New York 10523
Pergamon of Canada Ltd., 207 Queen's Quay West, Toronto 1
Pergamon Press (Aust.) Pty. Ltd., 19a Boundary Street,
Rushcutters Bay, N.S.W. 2011, Australia
Vieweg & Sohn GmbH, Burgplatz 1, Braunschweig

Copyright © 1972 A. K. Majumdar

All Rights Reserved. No part of this publication may be reproduced, stored in a retrieval system or transmitted, in any form or by any means, electronic, mechanical, photocopying, recording or otherwise, without the prior permission of Pergamon Press Ltd.

First edition 1972

Library of Congress Catalog Card No. 79–168613

Printed in Great Britain by A. Wheaton & Co., Exeter

08 016754 3

CONTENTS

PREFACE ix

1. Introduction 1

Organic Reagents in Inorganic Analysis 1
References 5

2. The Effect of Atomic Groupings and Substituents on Reagents 7

Cupferrons 7
Hydroxytriazenes 10
Hydroxamic Acids 13
Hydroxylamines 22
References 27

3. Preparation and Properties of *N*-Benzoylphenylhydroxylamine and its Analogues 30

N-*Benzoylphenylhydroxylamine (BPHA)* 30
N-*Substituted Phenylhydroxylamines* 33
N-*Acetylphenylhydroxylamine* 39
N-*2-Thiophenecarbonyl-p-tolylhydroxylamine* 40
Spectral Characteristics of p-*TTHA and PTHA* 41
N-*Benzoyl-o-tolylhydroxylamine* 41
N-*Benzoyl*-p-*tolylhydroxylamine and* N-*benzoyl*-m-*tolylhydroxylamine* 42
N-*Cinnamoylphenylhydroxylamine* 42
N-*Phenylacetylphenylhydroxylamine* 43
N-*Benzoyl*-p-*chlorophenylhydroxylamine* 43
N-*Salicylphenylhydroxylamine* 45
N-*Acetylsalicylphenylhydroxylamine* 46
N-*Benzoylmethylhydroxylamine* 46
N-*Substituted Arylhydroxylamines* 47
N-*Thiobenzoylphenylhydroxylamine (TBPHA)* 48
References 52

v

4. Gravimetric Determination of the Elements with N-Benzoylphenylhydroxylamine and its Analogues — 54

1. Applications of N-Benzoylphenylhydroxylamine — 55
2. Applications of N-Cinnamoylphenylhydroxylamine — 99
3. Applications of N-Benzoyl-o-tolylhydroxylamine — 100
4. Application of N-Salicylphenylhydroxylamine — 102
5. Application of N-Acetylsalicylphenylhydroxylamine — 102
6. Applications of Thiobenzoylphenylhydroxylamine — 103
References — 105

5. Spectrophotometric Determination of the Elements with N-Benzoylphenylhydroxylamine and its Analogues — 107

1. N-Benzoylphenylhydroxylamine as a Spectrophotometric Reagent — 109
2. N-Acetylphenylhydroxylamine as a Spectrophotometric Reagent — 134
3. N-Benzoylmethylhydroxylamine as a Spectrophotometric Reagent — 135
4. N-2-Thiophenecarbonyl-p-tolylhydroxylamine and N-2-thiophenecarbonylphenylhydroxylamine as Spectrophotometric Reagents — 136
5. N-Cinnamoylphenylhydroxylamine as a Spectrophotometric Reagent — 138
6. N-Benzoyl-p-tolylhydroxylamine as a Spectrophotometric Reagent — 141
7. N-Benzoyl-o-tolylhydroxylamine as a Spectrophotometric Reagent — 142
8. N-Benzoyl-p-chlorophenylhydroxylamine as a Spectrophotometric Reagent — 143
9. N-Benzoyl-o-tolylhydroxylamine and Other Aromatic Hydroxylamines as Spectrophotometric Reagents — 144
10. N-Furoylphenylhydroxylamine as a Spectrophotometric Reagent — 150
11. N-Acetylsalicylphenylhydroxylamine as a Spectrophotometric Reagent — 153
12. N-Arylhydroxylamines as Spectrophotometric Reagents — 154
References — 156

6. Separation of Elements by Solvent Extraction with N-Benzoylphenylhydroxylamine — 158

The Separation of Thorium and Uranium from Lanthanum — 159
Simultaneous Separation of Iron and Small Amounts of Titanium from Aluminium — 161
Separation of Scandium from the Lanthanides, Zirconium and Titanium — 162
Extraction of Aluminium from Complex Mixtures Including Uranium-based Fuels and Stainless Steels — 163
Extraction of Thorium and Separation from the Lanthanides — 164
Extraction of Tungsten — 164
Extraction of Zirconium — 165
Separation of Niobium from Tantalum — 166
The Separation of Niobium and Tantalum — 166

Extraction of Niobium, Tantalum, Titanium, Zirconium and Vanadium from Sulphuric Acid	167
Separation of Protactinium from Niobium, Titanium, Zirconium and Hafnium	169
Separation of Protactinium from Other Elements	169
Separation of Niobium from Zirconium	170
Separation of Tin and Antimony from Indium	171
Separation of Gallium, Indium, Thallium, Germanium, Tin and Lead	172
Separation of Protactinium from Niobium, Tantalum and Other Elements	173
Separation of Niobium and Tantalum Fluorocomplexes	176
Separation of Plutonium from Uranium, Americium, Zirconium and Other Fission Products	176
Separation of Protactinium from Neutron-irradiated Thorium	178
Separation of Arsenic, Antimony, Bismuth and Tin	179
Extraction of Thallium as a Function of pH and BPHA Concentration	180
Separation of Bismuth from Lead	180
Extraction of Copper, Iron, Lead, Nickel, Cobalt and Cadmium	181
Extraction of Zinc, Mercury, Bismuth, Manganese, Aluminium, Chromium and Silver	182
Extraction of Several Common Metal Ions	183
Extraction and Determination of Beryllium in Silicates	183
Separation of Elements from Hydrochloric Acid. Application to the Extraction of Niobium and Zirconium from Uranium	184
Separation of Cations by Extraction	186
Extraction of Germanium and Separation from Gallium	191
References	192

7. Titrimetric and Paper Chromatographic Applications of N-Benzoylphenylhydroxylamine — 193

Determination of Iron	194
Determination of Scandium and Zirconium	194
Determination of Vanadium(V) in the Presence of Fluoride, Phosphate, Titanium, Molybdenum and Manganese	194
Determination of Iron and Copper in the Presence of Many Other Ions by Using an Extractive End-point Procedure	195
Determination of Indium, Thallium and Thorium with Iron–BPHA as a Metallochromic Indicator	197
N-Benzoylphenylhydroxylamine as a Titrant for the Determination of Zirconium	198
Amperometric Determination of Titanium, Zirconium, Gallium, Scandium and Other Elements	198
Amperometric Determination of Gallium in Arsenides and Phosphides	201
Amperometric Determination of Niobium	201
Paper Chromatographic Separation of Metal Ions	202
References	204

INDEX   207

PREFACE

THE versatile complex forming ability of organic hydroxylamine derivatives such as N-benzoylphenylhydroxylamine (BPHA) and its analogues have interested me a great deal and have kept my colleagues and myself engaged for quite some time exploring the various analytical applications of these reagents. But never did it occur to me that I might write a monograph on the analytical applications of the organic hydroxylamine derivatives. I must, therefore, be thankful to Professor R. Belcher for kindly suggesting to me that I might undertake the work. It is really astonishing to observe that so much work has been done within a span of less than two decades. On going through the plethora of publications it is quite discernible that instead of a systematic and co-ordinated effort of adjudging the analytical potentialities of these reagents, simultaneous, many-pronged attempts have been made in various laboratories in the hope of finding some novel and extraordinary analytical applications. This, no doubt, unearthed a wealth of information, but at the same time has necessitated the processing of all the available data in the form of a systematic, comprehensive and well-documented account.

The introductory chapter, which starts with a brief historical account of the uses of organic reagents in inorganic analysis, is essentially devoted to the discussion of some basic physico-chemical factors intimately associated with the constitution of a good reagent. This includes some fundamental ideas about the type and nature of the principal reactive and other groupings in the ligand, the stability of the complex and the factors responsible for the solubility, selectivity and sensitivity of the reagent and steric effects. This and the subsequent chapter dealing with the effects of atomic groupings and substituents on the applicability of the reagents to various aspects of inorganic analysis should reveal at a glance the primary reasons for the analytical potential of the organic hydroxylamine derivatives. To press this point further, a few important analytical applications of reagents having atomic groupings analogous to those of the hydroxylamine derivatives have also been included.

The subject proper begins in the third chapter with the inclusion of the methods of preparation and properties of BPHA and its analogues and derivatives, and their reactions with different metal ions.

The four subsequent chapters deal with the various types of analytical applications of BPHA and its analogues which include gravimetry, spectrophotometry, solvent extraction and also their applications in titrimetry and paper chromatography. An attempt has been made to bring the information reported by various authors in many types of journal and using different modes of presentation into some kind of standardized form. This necessitated in some instances a re-examination in the laboratory of the methods of preparation of the reagents and of the different analytical procedures, to bring about some uniformity in the mode of presentation. Observations are given as footnotes under the respective methods.

I must accept the primary responsibility for those errors and omissions which, in spite of all the best efforts and precautions, have managed to creep into the manuscript.

I express my thanks to my colleagues and co-workers, Mr. A. K. Chakraburtty, Mr. S. K. Bhowal, Mr. B. K. Mitra, Dr. S. P. Bag, Dr. B. K. Pal, Dr. A. B. Chatterjee, Dr. Gayatri Das and Mr. S. Lahiry, who helped in the preparation of the manuscript.

Finally, the author wishes to express his sincere gratitude to his wife for her constant encouragement and patient co-operation.

Procedures in some cases, especially where details were not available, were worked out and incorporated.

A. K. M.

CHAPTER 1

INTRODUCTION

Organic Reagents in Inorganic Analysis

AS AN introduction to the subject matter of the monograph, the application of the organic bidentate ligand N-benzoylphenylhydroxylamine and its analogues, it is necessary to have a brief discussion on the nature, classification and behaviour of organic reagents, the specificity, selectivity, sensitivity of their reactions with inorganic ions and the solubility and stability of the products of their reactions. These are some of the factors which are taken into account in the evaluation of analytical methods involving organic reagents which, by their ease of substitution, afford reagents of desired analytical properties.

Ever since Pliny's observation[1] about the use of gall-nut tannin for producing a blue colour with iron, and Varahamihira in the sixth century A.D. in India alluded to the preparation of fast dyes from natural dyes like *manjishtha* (madder) with alum and other chemicals,[2] the use of organic compounds as reagents has developed as a chemical science in itself. Wiedemann's observation (1848) that the copper(II) colour reaction with biuret was due to the amide linkage, and that of Ilinski and von Knorre of α-nitroso-β-naphthol as a reagent[3] for cobalt and iron(III) have been followed by systematic researches resulting in the accumulation of facts, vast and varied, which has been dealt with in many treatises and monographs.[4–12]

Organic reagents, according to their reactions with inorganic ions, are classified under four broad categories, viz.

(i) normal salt-forming types, such as phenylarsonic acid, which precipitates quantitatively zirconium, thorium,[13] tin,[14] bismuth,[15-18] lead,[19] niobium and tantalum[20,21] and permits satisfactory separation of niobium and tantalum from each other;[22]

(ii) reagents that form adsorption compounds, such as those formed with tannin, which is used particularly for the precipitation of niobium and tantalum;[23]

(iii) uncharged organic molecules that form metal chelates, such as the coloured complexes formed by iron(II) with 2,2′-dipyridyl and 1,10-phenanthroline;[24]

(iv) reagents that form uncharged metal chelates (inner complexes).

Inner complexes in which the metal ions are bound to the ligands via at least two donor atoms must possess at least one acid group with a replaceable hydrogen atom and a basic, coordinating donor group.

Salt-forming groups usually are —CO_2H (carboxyl), —SO_2H (sulphinic), —SO_3H (sulphonic), —OH (enolic or phenolic hydroxyl), =N(H)(O) or =N—OH (oxime), —NH—H (primary amino), —NR—H (secondary amino), —NO(OH) (nitroxyl), —SH (mercapto), —$AsO(OH)_2$ (arsonic) and —$As(OH)_2$ (arsinic). Donor groups with oxygen, sulphur and nitrogen atoms as centres of coordination consist of —NH_2 (primary amine), —NHR (secondary amine), —NR_2 (tertiary amine), =NH (imino), =NR (substituted imino), —NO (nitroso), —NO_2 (nitro), =NOH (oxime), —N=N—(azo), ≩N (ring N), —OH (alcoholic hydroxyl), ⊃O (ether), ⊃S (thio ether), =CS (thiocarbonyl) and =CO (carbonyl). With alkyl or aryl substitution while the donor power of a sulphur atom increases that of nitrogen or oxygen diminishes. Substitution of an electronegative group in the molecule increases the acidity and decreases the basicity of acid and basic groups, respectively. Coordinating ability of oxygen and nitrogen in the most common donor groups are in the order.[25]

$-O^-$	>	COO^-	>	$-O-$	>	$C=O$
enolate ion		carboxylate		ether		carbonyl

$-NH_2$	>	$\geqslant N$	>	$-N=N-$	>	$\equiv N$
amine		ring N		azo		triple bonded N

For analytical applications, reagents are usually so chosen that their acid and basic groups are in such positions as to form five- or six-membered chelate rings.

There is no known specific reagent. However, by the right choice of masking agents and the judicious conditioning of the reaction environment, the reactions in which a reagent can participate can be made very selective or even specific. Moreover, selectivity of reactions often increases as the functional groups which form chelates through two oxygen atoms are changed to those chelating through one oxygen and one nitrogen atom and further when chelation is through two nitrogen atoms.[26,27] Even greater selectivity is expected when the electronegativity of the coordinating atom is increased. Also such an electronegative group may enhance the acidity and thus make the reagent more selective as the complexes formed thereof are less stable.[28,29] Steric factors also govern the stability of chelates and hence the selectivity of reactions.

Extraction of a coloured reaction product into an immiscible organic solvent is very often used to increase the sensitivity. Even soluble organic liquids which change the activities of the ions in a colour reaction increase the sensitivity. The weighting effect increases the sensitivity of a precipitation reaction.

Water solubility depends on whether the molecule contains any water-solubilizing group, i.e. a group which can be hydrated. Since an acidic hydrogen atom is a hydration centre, due to its hydrogen bonding with a water molecule, the presence of groups such as $-OH$, $-CO_2H$, $-SO_3H$ in the organic molecule promotes water solubility. The presence of $-NR_2$ groups (R=H or alkyl) also increases water solubility, as the nitrogen atom is an active centre for binding water molecules.

Constitutional factors also have a role. 1,2,3-Trihydroxybenzene is fairly soluble in water while 1,3,5-trihydroxybenzene is only slightly so. Again, if the —OH group capable of hydrogen bond formation is present at a position ortho to a coordinating centre such as =CO, —NO$_2$ or —N=N—, the molecule is less soluble than the para- and meta-isomers because of the formation of a ring owing to hydrogen bonding.[30,31]

The stability of complexes with ionic or multidentate ligands is expected to increase as the ionic potential of the metal ion increases. This is borne out from the observed stability sequences of complexes of the typical non-transition metal ions within their own family and with same charge. The B-family members with a pseudo inert gas configuration form more stable complexes than those of the A-family members with an inert gas configuration, even though they are of the same size and charge, because of the former's high effective nuclear charge. For the transition metal ions, with incomplete d-orbitals, the stability also depends on the number of d-electrons, i.e. on the crystal field stabilization energy, electron pairing energy, stereochemical configuration and on metal to ligand and ligand to metal π-bonding.

The metal ions are classified[32,33] as class (a) and class (b) ions according to whether they form, respectively, their more stable complexes with the ligand atoms of the first periodic row, i.e. N, O and F, or with those of the second or subsequent rows. The class (a) or hard acid character of the acceptor atom is associated with small size, high positive charge and absence of valence electrons which are easily distorted, while the class (b) or soft acid character is manifested by the acceptor atom of low or zero charge, of large size or having several valence electrons which are easily distorted. The stability of the class (b) metal ion complexes decreases in the order of the increasing electronegativity of the ligand atoms as C~S > I > Br > Cl~N > O > F. A strongly reversed trend, although not complete, is observed with class (a) metal complexes.

The simple electrostatic approach explains the stability of many of the complexes of the metal ions, particularly of the class (a) type. For class (b) metal ions, besides electrostatic contribution, other factors such as π-bonding, crystal field effect and increased covalent character in metal ligand bond also play significant parts.

According to π-bonding theory, in addition to σ-bonding, class (a) metal ions have the potential to form ligand to metal π-bonds with the more basic ligand atoms, whereas the class (b) metal ions are likely to form metal to ligand π-bonds by the overlap of filled d-orbitals with the vacant p or d orbitals of the ligand. In the latter instance, the metal to ligand π-bonding increases the d-orbital splitting as the metal ion acquires a higher effective charge. This property of the ligand increases the crystal field stabilization energy compared to the ligands which have no such π-bonding tendency.

The stability of metal complexes, as well as depending on the type of metal ion and the donor atoms, rests much on the base strength and the number of coordination sites of the ligand, the chelate ring size, and steric and the resonance effects.

Complexes with five- and six-membered chelate rings are most stable; chelate complexes with four-membered rings are rare while chelate rings with more than six members are generally less stable. Of the former two, aliphatic chelate complexes with five-membered rings appear to be more stable owing to favourable entropy changes.[34] The six-membered chelate rings with conjugated double bonds or aromatic ligands are sometimes more stable compared to the five-membered rings, because perhaps of the release in strain by wider bond angles and resonance. The stability decreases with the decrease in the double bond order of the chelate ring.[35]

Steric inhibition of complex formation may be due to several factors, viz. (i) the presence of a bulky group attached to or close to the coordinating atom and (ii) lack of basic geometry for the overlap of the metal orbitals with the ligand orbitals, a prerequisite for the formation of a stable structure.

References

1. NIERENSTEIN, M., *Analyst* **68**, 212 (1943).
2. RAY, P., *History of Chemistry in Ancient and Mediaeval India*, Indian Chemical Society, Calcutta, 1956, p. 103.
3. ILINSKI, M. and VON KNORRE, G., *Ber.* **18**, 699 (1885).
4. DIEHL, H., *Chem. Rev.* **21**, 39 (1937).

5. MELLON, I., *Organic Reagents in Inorganic Analysis*, Blakiston Co., Philadelphia, 1941.
6. PRODINGER, W., *Organic Reagents Used in Quantitative Inorganic Analysis*, Elsevier Publishing Co., New York, 1940.
7. YOE, J. H. and SARVER, L. A., *Organic Analytical Reagents*, John Wiley & Sons, New York, 1941.
8. WELCHER, F. J., *Organic Analytical Reagents*, Vols. 1–4, D. Van Nostrand Co., New York, 1947–8.
9. FLAGG, J. F., *Organic Reagents Used in Gravimetric and Volumetric Analysis*, Interscience Publishers, New York, 1948.
10. FEIGL, F., *Ind. Eng. Chem. Anal. Ed.* **8**, 401 (1936).
11. FEIGL, F., *Chemistry of Specific, Selective and Sensitive Reactions*, Academic Press, New York, 1949.
12. MAY, I. and SCHUBERT, L., "Reactive Groups as Reagents, Inorganic Applications," in KOLTHOFF, I. M. and ELVING, P. J. (Eds.), *Treatise on Analytical Chemistry*, Part I, Vol. 2, Interscience Publishers, New York, 1961.
13. RICE, A. C., FOGG, H. C. and JAMES, C., *J. Am. Chem. Soc.* **48**, 895 (1926).
14. KNAPPER, J. S., CRAIG, K. A. and CHANDLEE, G. C., *J. Am. Chem. Soc.* **55**, 3945 (1933).
15. MAJUMDAR, A. K., *J. Indian Chem. Soc.* **21**, 119 (1944).
16. MAJUMDAR, A. K., *J. Indian Chem. Soc.* **21**, 188 (1944).
17. MAJUMDAR, A. K., *J. Indian Chem. Soc.* **22**, 313 (1945).
18. MAJUMDAR, A. K. and SEN SARMA, R. N., *J. Indian Chem. Soc.* **26**, 477 (1949).
19. MAJUMDAR, A. K. and SEN SARMA, R. N., *J. Indian Chem. Soc.* **28**, 654 (1951).
20. ALIMARIN, I. P. and FRID, B. I., *Mikrochemie* **23**, 17 (1937).
21. RANKAMA, K., *Bull. Comm. Geol. Finlande* No. 133 (1944).
22. MAJUMDAR, A. K. and MUKHERJEE, A. K., *Anal. Chim. Acta* **21**, 330 (1959).
23. SCHOELLER, W. R., *The Analytical Chemistry of Tantalum and Niobium*, Chapman & Hall, London, 1937.
24. BLAU, F., *Monatsch* **19**, 647 (1898).
25. VAN UITERT, L. G. and FERNELIUS, W. C., *J. Am. Chem. Soc.* **76**, 379 (1954).
26. IRVING, H. and WILLIAMS, R. J. P., *Analyst* **77**, 813 (1952).
27. IZATT, R. M., FERNELIUS, W. C. and BLOCK, B. P., *J. Phys. Chem.* **59**, 235 (1955).
28. REID, J. C. and CALVIN, M., *J. Am. Chem. Soc.* **72**, 2948 (1950).
29. IRVING, H., HOLLINGSHEAD, R. G. W. and HARRIS, G. *Analyst* **80**, 260 (1955).
30. SIDGWICK, N. V. and CALLOW, R. K., *J. Chem. Soc.* **125**, 527 (1924).
31. SIDGWICK, N. V., *J. Chem. Soc.* **127**, 907 (1925).
32. AHRLAND, S., CHATT, J. and DAVIES, N. R., *Quart. Rev. London* **12**, 265 (1958).
33. PEARSON, R. G., *J. Am. Chem. Soc.* **85**, 3533 (1963).
34. COTTON, F. A. and HARRIS, F. E., *J. Phys. Chem.* **59**, 1203 (1955).
35. CALVIN, M. and WILSON, K. W., *J. Am. Chem. Soc.* **67**, 2003 (1945).

CHAPTER 2

THE EFFECT OF ATOMIC GROUPINGS AND SUBSTITUENTS ON REAGENTS

OF THE substituted hydroxylamines, the ammonium salt of *N*-nitrosophenylhydroxylamine and *N*-benzoylphenylhydroxylamine are the two which have found wide application as analytical reagents. Although the reports describing the uses of the others are scanty, some of these reagents are sensitive and versatile. The following is a brief account of the reactions of these hydroxylamines with metal ions according to their reactive atomic groupings, so as to describe and discuss the effects of substituents on these reactions.

Cupferrons

The ammonium salt of *N*-nitrosophenylhydroxylamine has the trivial name of cupferron, because it produces precipitates with copper and iron[1] in strong mineral acid solution. The reagent is

Cupferron

readily soluble in water, alcohol, ethyl acetate and chloroform. It is, however, unstable to visible or ultraviolet light, and to air oxidation. Because it decomposes in solution, especially on heating, to nitrosobenzene, a freshly prepared solution must always be used as the precipitant and the precipitation done in the cold. The complexes lack that stability which is expected of a salt of the inner complex type and hence the cupferron precipitates are ignited to oxides before their final weighing. However, by precipitation from homogeneous solution, a copper complex of definite composition, free from decomposition products and which can be weighed after drying, is obtained.[2]

The ligand is considered to be bidentate, with coordination to the oxygen of the nitroso group and to the negatively charged oxygen. The acidity of the free acid derived from cupferron is due to the presence of the electronegative —NO group and the phenyl group. Metal ions with high charge coordinate to oxygen and hence group IV A metals with small size and high charge show especial reactivity towards cupferron. Substitution in the phenyl ring does not change its precipitating character.

As an analytical reagent, cupferron has little selectivity, but the selectivity is somewhat greater in strongly acidic solutions than in weakly acidic solutions. Under the former conditions, only vanadium, titanium, tin, gallium, iron(III), niobium, tantalum, zirconium and presumably hafnium precipitate.[3] As the acidity of the solution is reduced other ions are precipitated. Among the interesting separations that can be achieved at high acidity mention can be made of the separation of uranium(IV) from uranium(VI), of vanadium(V) from uranium(VI) and of vanadium from phosphorus. Tartaric, oxalic and fluoboric acids do not affect cupferron precipitation. In the presence of tartaric acid, quantitative separation of niobium from tantalum, when present in a ratio of 1:30 to 30:1, has been effected at pH 4·5–5·5 using tin(II) or tin(IV) as co-precipitant.[4] The selectivity of cupferron precipitation at higher pH values is increased to a marked degree by EDTA. Thus, titanium and uranium(VI) are separated from a large selection of ions.[5,6] Selectivity for titanium is further increased by the solvent extraction of its complex into 4-methyl-2-pentanone for its spectrophotometric determination.[7]

Most of the metal complexes of cupferron are soluble in chloroform, ethyl acetate, ether, methyl isobutyl ketone, benzene and o-dichlorobenzene. Chloroform is the preferred solvent. The solubility of some of the complexes in chloroform and ether is used for their separation at different acidities.[8-10]

A nephelometric determination of vanadium with cupferron, using gum arabic to stabilize the suspension, has been proposed.[11] An amperometric titration of titanium, zirconium and hafnium can be undertaken in 10% sulphuric acid under which conditions the reaction is stoichiometric.[12] Satisfactory results are obtained even in presence of phosphate.

Neocupferron, the ammonium salt of N-nitrosonaphthylhydroxylamine, is more stable than cupferron in aqueous solution, but still decomposes on standing for a long period. Another interesting point about the reagent is that its neodymium salt is extractable[13] in organic solvents. It is similar to cupferron in its analytical reactions, but owing to the weighting effect of the naphthyl residue, its copper and iron precipitates are less soluble and more voluminous. Thus it makes possible the determination of these elements in mineral water without pre-concentration.

The 2-fluorenyl analogue of cupferron has many characteristics in common with cupferron and in its reaction with iron it is as sensitive as neocupferron but seems to have no advantage over the latter as an analytical reagent.[14] Iron to the extent of 5 ppm can be determined in hydrochloric or acetic acid, but not in sulphuric acid. Copper in higher concentrations interferes. The precipitates of iron(III), antimony, cobalt and lead are soluble in chloroform.

Studies on the spectrophotometric, polarographic and solubility behaviour of cupferron, neocupferron and p-phenyl cupferron (p-xenyl analogue) chelates of group IV B metals indicate that in 4 M perchloric acid, titanium, zirconium and hafnium form water-soluble bis chelates MA_2^{2+}. From their half-wave potentials, it appears that the titanium chelates are least stable, whereas hafnium produces the most stable chelates.[15] The molar absorptivities of the chloroform extracts of cupferron and neocupferron chelates are approximately equal, but those of the p-xenyl compounds are much higher. For each series, that of zirconium is the least.

Hydroxytriazenes

The hydroxyphenyl- and hydroxymethyltriazenes, which have the functional grouping —N(OH)—N=N—, are mostly stable to heat, light and air-oxidation and are soluble in many organic solvents. Their metal complexes are also soluble in many such solvents from which they can be crystallized with definite melting points. The hydroxymethyltriazenes are more soluble in water than the hydroxyphenyltriazenes, which are sparingly soluble or insoluble in water unless a solubilizing group is present in the molecule.

For the hydroxytriazenes, an intramolecular hydrogen-bonded structure has been proposed from ultraviolet and infra-red absorption studies. 3-Hydroxy-1,3-diphenyltriazene has been suggested to have tautomeric forms, on the basis of its ultraviolet spectrum and acid dissociation constant.[16]

3-Hydroxy-1,3-diphenyltriazene is susceptible to acid hydrolysis.[17] Even in weakly acidic solution it hydrolyses to give water-soluble products. At a pH below 3, copper(II), palladium(II), titanium(IV), iron(II), iron(III), vanadium(III), vanadium(V) and molybdenum(VI) complexes are precipitated. But apart from precipitates with copper and palladium, they are decomposed on heating in the acidic medium. Thus, 3-hydroxy-1,3-diphenyltriazene is selective for the gravimetric determination of copper and palladium by direct weighing as $M(C_{12}H_{10}N_3O)_2$, with $M = Cu$ or Pd.

The yellowish-brown palladium and chocolate-coloured copper complexes are soluble in many organic solvents, and can be crystallized from acetone as silky needles with melting points 193° and 252°, respectively. The nickel complex which precipitates at pH 4·4–7·0 has similar properties and after crystallization from acetone melts at 209°. The titanium(IV) complex is appreciably soluble in ethanol, but decomposes when the temperature of the medium from which it precipitates is raised to 55°. The titanium(IV) complex of 3-hydroxy-1-p-chlorophenyl-3-phenyltriazene, on the other hand, is stable in aqueous medium even at 100°. However, the precipitate is not of definite composition and has to be ignited to titania as the weighing form.[18] The optimum pH range for its complete precipitation is very narrow, pH 2·2–2·5, as at a higher pH it is partly hydrolysed and at a lower pH slight decomposition takes place. The bright-orange complex is readily soluble in chloroform, benzene and acetone, but less so in methanol and ethanol. It crystallizes from ethanol as bright-yellow needles, melting point 146°.

The complexes formed by the *p*-chloro derivative are more stable, no doubt, than those formed by the parent triazene but the compound offers no special advantage in the determination of copper, palladium or nickel. The complexes formed of the *o*-chloro isomer are not granular and are comparably less stable. The steric effect of the chlorine atom ortho to the azo nitrogen may be the reason for the lower stability. On the other hand, the higher stability of the complexes prepared from 3-hydroxy-1-*p*-chlorophenyl-3-phenyltriazene may be due to resonance stabilization effected by the development of the positive charge, due to coordination, on the nitrogen atom para to chlorine.

By the substitution of the *o*-chloro group with a carboxyl group, 1-(*o*-carboxyphenyl)-3-hydroxy-3-phenyltriazene is produced. It is stable in acidic solutions and readily soluble in ammoniacal and alkaline solutions. Its titanium(IV) complex is also highly stable in acidic media even at boiling temperature but it hydrolyses at and above pH 6. The complexes of titanium(IV), vanadium(V), palladium(II) and copper(II) are all soluble in many organic liquids including pyridine and chloroform. The iron(III) complex, though highly soluble in pyridine but less so in alcohols, is almost insoluble in all other solvents unless a drop of ammonia or alkaline solution is added. On crystallization from a chloroform–nitrobenzene mixture, the orange-red crystals of the titanium complex melt with decomposition at 283–4°, whereas the crystals of the vanadium(V), iron(III), palladium(II) and copper(II) complexes decompose at 165°, 212°, 250° and 305°, respectively. The palladium and copper complexes precipitated at pH 3–6 and 3·0–5·5, respectively, are of composition $M(C_{13}H_9O_3N_3) \cdot H_2O$. The nickel and cobalt complexes are somewhat soluble in the reaction medium.

This triazene derivative is an excellent, highly selective gravimetric reagent[19] for titanium when EDTA is present to mask other ions. The titanium compound precipitated at pH 2–5 is $TiO(C_{13}H_{10}O_3N_3)_2$. The derivative also behaves as a spectrophotometric reagent for the determination of iron, titanium and vanadium in presence of each other and other ions.[20] The coloured species due to these three ions on extraction in chloroform at pH 9·5, 2·0 and 2·0, respectively, show absorption maxima at about 500, 430 and 410 nm, with sensitivities 0·01126, 0·0038 and 0·0062 μg cm^{-2}, respectively. A study of the composition of the coloured extracted species by Job's method indicates the formation by iron(III) and titanium(IV) of 1:2 chelates and by vanadium(V) of a 1:1 chelate with the triazene. The iron (III) and vanadium (V) complexes have the compositions FeH$(C_{13}H_9O_3N_3)_2$ and $V_2O_3(C_{13}H_9O_3N_3)_2$, respectively.

From a study of the infra-red spectra of the isolated chelates, it appears that whereas vanadium, iron, palladium and copper replace the hydrogen atoms of the salt forming —OH *and* —COOH groups, titanium replaces only that of the —OH group. The presence of a free carboxyl group in the titanium(IV) salt is further confirmed by the fact that it is soluble in water at and above pH 6, before hydrolysis occurs, and that the methyl ester of the reagent gives the same titanium reaction. Thus this triazene is a bidentate ligand towards titanium, but acts as a tridentate one in its reactions with some other ions.

The red compound, 3-hydroxy-1-(*o*-nitrophenyl)-3-phenyltriazene, gives a red precipitate with zinc at pH 5–6 (dilution limit, 1:3 × 10^4) and a violet precipitate with nickel (dilution limit, 1:5 × 10^4). It has been used for the colorimetric determination of zinc.[16]

3-Hydroxy-1-(*p*-nitrophenyl)-3-phenyltriazene, on the other hand, is a greenish-yellow product, which is insoluble in water but soluble in ethanol, acetone and toluene, and dissolves in alkaline solution to give, finally, a red colour. It finds application as a colorimetric reagent for magnesium, which in neutral or weakly acidic solution develops a rose-violet or blue-violet colour, depending on the concentration, with a detection limit of 1 ppm. Amounts of magnesium greater than 0·05 mg per ml produce a precipitate.[16]

3-Hydroxy-1-*p*-tolyl-3-phenyltriazene[21] and 3-hydroxy-1,3-*p*-ditolyltriazene[16] have also been studied for the direct gravimetric determination of palladium and copper. These compounds have more favourable conversion factors compared to 3-hydroxy-1,3-diphenyltriazene.

3-Hydroxy-1-(*p*-sulphophenyl)-3-phenyltriazene sodium salt, is a water-soluble compound that produces water-soluble complexes. The solid starts to decompose at 157° and has a pK_a value of 9·99. It has been used mainly as a colorimetric reagent for the determinations[16] of 0·2–5·0 ppm of palladium (430 nm, pH 2·5–3·0), 1–20 ppm of iron(III) (pH 2·0–4·3, λ_{max} 520 and 650 nm), 1·5–25·0 ppm of molybdenum(VI) (pH 2·8–3·7, λ_{max} 416 nm), 0·3–2·5 ppm of titanium(IV) (pH 1·9–3·3, at 425 nm), copper(II) (pH 4·4–6·2, λ_{max} 405 nm), cobalt(II) (pH 4·3–5·9, λ_{max} 390 nm, plateau from 450–485 nm) and nickel(II) (pH 3·8–5·9, λ_{max} 420 nm). It is also used as an indicator for the complexometric titration of iron(III), and for the indirect spectrophotometric determination of fluoride, phosphate, oxalate, citrate and tartrate, based on the masking of the colour of the iron(III) complex by these ions.

3-Hydroxy-1-phenyl-3-methyltriazene and 3-hydroxy-1-*p*-tolyl-3-methyltriazene have also been suggested as reagents, the former for the spectrophotometric determination[16] of 1–10 ppm of iron(III) (pH 3·1–4·5, 625 nm) and the latter for the gravimetric determination[22] of palladium and copper by precipitation at pH 1·2–3·0 and 3·3–4·0 respectively. The former is freely soluble in water–alcohol mixtures, and the latter is fairly soluble in hot water.

Hydroxamic Acids

Organic hydroxamic acids may be classified as oximes, with two tautomeric forms, (I) and (II), with the keto form (I) predominating

(I)   (II)   (III)

in acidic solutions. Because the *N*-substituted hydroxamic acids, give similar reactions with metal ions to the unsubstituted acids

$$\underset{\underset{\text{R}}{}}{}\text{R}-\underset{\|}{\overset{\text{O}}{\text{C}}}-\underset{|}{\overset{\text{OH}}{\text{N}}}-\text{H}$$

the metal complexes are considered to have the structure (III). That the hydroxamic acid-metal ion complexes with the latter group of reagents are all soluble in alkaline solutions clearly shows that the acidic N-hydrogen does not take part in complex formation in an acidic medium. Moreover, the vanadium benzohydroxamic acid complex, formed in acidic solution, is readily extracted in organic solvents,[23,24] whereas that formed in solution at a high pH cannot be so extracted.[24] Presumably, the former is uncharged and the latter is anionic,[23,24] derived from the enol form (II) of the hydroxamic acid. Cobalt(III), manganese(III) and iron(III) complexes of the type $[M^{III} R_3]^{3-}$, where RH_2 is a molecule of a hydroxamic acid, have been isolated as compounds with the cobalt(III) tris-biguanidinium cation.[25] The anionic nature of the manganese(III) complexes has also been verified by electrochromatographic migration[25] and adsorption by an anion exchange resin.[26]

The extractability of hydroxamic acid chelates depends on the nature of the reagent.[23] Hydroxamic acids with aromatic substituents favour an organic solvent (hexanol) while those with aliphatic substituents prefer the aqueous phase. The colours produced by vanadium(V) with the aromatic acids are comparable to that with benzohydroxamic acid, but the aliphatic acids ordinarily either impart practically no colour, as with the aceto and propiono derivatives, or only a colour of very low intensity. This arises because charge transfer between the aromatic ligand and the metal ion gives rise to intense colours.

In an attempt to correlate structure with the colour reaction of various hydroxamic acids with vanadate, it was observed[24] that, of the aliphatic acids, sorbohydroxamic acid with a conjugated double bond system gave a more intense colour than those with saturated carbon atoms near the reactive group. Substituted aliphatic hydroxamic acids such as glycinohydroxamic acid, cyclohexanone-bis (carbohydroxamic acid) ketal, chlor-, dichlor- and trichloracetohydroxamic acids were unreactive towards vanadium. The non-availability of electrons for complex bonding owing to the presence of strongly electronegative chlorine or oxygen atoms near the reactive grouping is probably the reason for their such non-reactive character. In glycinohydroxamic acid, however, the interpositioning of a —CH_2— group reduces the electron withdrawing effect enough to make the acid react with nickel, copper, cobalt(II), iron(III), osmium(VI) and uranium(VI) ions to produce colours or precipitates.

Of the four aromatic hydroxamic acids investigated, 1-naphthohydroxamic acid was least reactive; 2-naphthohydroxamic acid gave a very intense colour but those with benzo- and 1-naphtho-hydroxamic acids were much less intense, and very similar to each other. 1-Naphthoacethydroxamic acid shows a much decreased absorption, owing to the insulating effect of the methylene group placed between the reactive group and the aromatic nucleus. Substituted aromatic hydroxamic acids give colour reactions with intensities in the order: *p*-(diphenylmethyl)benzohydroxamic acid > salicylhydroxamic acid > *p*-toluohydroxamic acid ≃ *p*-chlorobenzohydroxamic acid > *p*-nitrobenzohydroxamic acid > *m*-toluohydroxamic acid > 2,3-dichlorobenzohydroxamic acid. The last, with two chlorine atoms, is the least reactive, yielding colours only with vanadium(V) and iron(III), whereas with others of this class, reactions with vanadium(V), iron(III), uranium(VI) and copper (II) are more pronounced than those with titanium(III), thallium(III) and osmium(VI).

Heterocyclic compounds such as the pyridine derivatives of hydroxamic acids do not give any colour with vanadium. Nicotino- and isonicotino-hydroxamic acids, however, give colours of low intensity, which may arise out of oxidation-reduction reactions.

C-Substituted hydroxamic acids are detected by their colour reaction with iron(III);[27] the most characteristic reaction of these hydroxamic acids is to impart, in acidic medium, a violet-red colour which changes to wine-red and thence to orange-yellow as the pH is raised. The composition of the complexes in solution appears to be 1:1 (Fe:acid) below pH 2, 1:2 at about pH 3 and 1:3 above pH 5. The last type tends to be anionic. The first two types are extremely water-soluble and the last type, particularly for aliphatic hydroxamic acids, though very soluble, can be isolated in cases like acetohydroxamic acid by the use of organic solvents. The solubility of the latter type is considerably reduced by the introduction of a group with weighting or size effects such as phenyl.[25] These acids have all the potentialities of gravimetric reagents. As well as weighing, determinations involving hydroxamate precipitates may be completed in other ways.[28,29] The precipitate may either be dissolved in dilute acetic acid and the hydroxamic acid liberated determined colorimetrically with iron(III) chloride or it may be digested with hydrochloric acid to produce hydroxylamine, which is reduced by titanium(III) chloride and the excess of the latter back titrated with iron(III). As precipitants, these acids are non-selective, but by the control of reaction conditions they can be rendered somewhat selective in their action. Benzo- and salicylhydroxamic acids, first introduced to explore the possibilities of the gravimetric determination of copper, cobalt and nickel,[30] find many specific applications which deserve special mention.

BENZOHYDROXAMIC ACID

Benzohydroxamic acid, $C_6H_5 \cdot CO \cdot NHOH$, is a bidentate chelating agent that loses one proton on forming a chelate.[31] From the ultra-violet absorption spectrum[32] and infra-red analysis,[33] it is inferred that the acid exists mainly in the keto form, but in basic solution it changes considerably to the enol form. The benzohydroxamic acid anion has been suggested to have contributions from A and from either or both of the forms B or/and C. From kinetic studies, however, the A and C forms are proved to exist in approximately equivalent amounts.[34]

Benzohydroxamic acid is difficultly soluble in water. Its water-soluble potassium salt decomposes quickly in basic solution. The red-brown iron(III) complex, $Fe(C_7H_6O_2N)_3$, is sparingly soluble in water, but dissolves in acids and alkalis and also in most organic solvents. Copper at pH 3–6 and nickel at pH 6 produce, respectively, bluish-green and greenish-yellow compounds of composition MR_2 (RH is a molecule of hydroxamic acid). The light pink cobalt complex changes in alkaline solution to a reddish-brown, anionic cobalt(III) species.[25] The manganese(II) complex, which is pale yellow or off-white, also changes as the pH of the solution is raised to above 9, to an intensely purple solution. The reaction is accelerated by traces of hydrogen peroxide. This change is due to the oxidation of manganese(II) to manganese(III).[25,26] Uranium(VI), molybdenum(VI), zirconium, titanium(IV) and vanadium(V) form oxy salts.[25]

Depending on the pH, iron forms three different complexes[23,25] in solution. A purple-red 1:1 (iron:acid) product at or below pH 1, a red 1:2 species at about pH 3·5 and at pH 8, another yellow-red 1:3 product has been identified having λ_{max} at 510, 480 and 440 nm, respectively. As the concentration of iron is increased, precipitates are formed at or above pH 4·7 which are all of the same composition, $Fe(C_7H_6O_2N)_3 \cdot 3H_2O$. An alcohol or ester extract of the coloured species formed at pH 2–9 has an identical absorption spectrum which indicates the singular nature of the extracted species. Alcohols increase the colour intensity, which is least sensitive to the effect of pH at pH 6–9.

The reaction of vanadium, like iron, is pH dependent.[23,35] Below pH 5, vanadium(V) combines with benzohydroxamic acid in aqueous solution in a ratio of 1:2, whereas above pH 7 it forms a 1:3 species and at pH 8 the complex is anionic, so that it is not soluble in organic solvents. In 50% ethanol, a 1:1 complex is formed at an apparent pH of 2·7. The complex isolated at pH 2 has the formula $H[VO_2R_2] \cdot H_2O$. It is extractable into polar oxygenated solvents such as alcohols, ketones, phthalate esters and trialkylphosphates, but not into hydrocarbons or ethers.[36] Alcohols enhance the colour intensity. Hence methods have been developed for the colorimetric determination of the element either at 500 nm in water–alcohol mixtures at pH 2·7[35] or at 450 nm in 1-hexanol extracts from solutions at pH 2.[36] The colour reaction is so highly sensitive below pH 3 that it detects even 10^{-8} g of vanadium;[35] the limiting dilution is $1:5 \times 10^6$.

The orange product formed with uranium in ammoniacal solution detects[35] on peptization the presence of uranium at a dilution $1:1 \times 10^6$. At all pH values soluble complexes are formed with *ca.* 1 mg of uranium. As the pH of the solution is raised from 3–4 to 5, the colour changes from red to red-orange and at 7 it is orange. Above pH 11, it diminishes. Uranium combines with the acid to form a 1:2 complex in solutions above pH 6, while at pH 4, the ratio is 1:1. However, from solutions of higher uranium concentrations a precipitate is obtained, of composition $UO_2R_2 \cdot H_2O$.[37] On extraction into polar oxygenated solvents, uranium yields the same complex at all pH values between 4 and 10, as shown by the absorption spectra. For its colorimetric determination, the uranium complex is extracted in 1-hexanol at pH 6·2 before its absorbance is measured at 380 nm.

The colour intensity of the anionic manganese complex formed above pH 10 is the basis of a method for the photometric determination of manganese[26] at 500 nm. The limiting dilution for the detection of manganese is $1:10^6$.[35]

Benzohydroxamic acid has been applied effectively for the separation at pH 4·0–6·4 of niobium and tantalum from each other. Tantalum precipitates completely from an oxalate solution on digestion.[38] Titanium and zirconium compounds, $TiO(C_7H_6O_2N)_2$ and $ZrO(C_7H_6O_2N)_2$, formed between pH 2·5 and 6, can be separated and weighed as such.[39]

As a complexing agent, the acid helps in the preferential extraction of americium in chloroform to separate it from actinium and some rare-earth elements.[40]

Benzohydroxamic acid has also been used as an indicator in the titration of EDTA with iron(III).[41]

SALICYLHYDROXAMIC ACID

The salicylhydroxamic acid molecule possesses, apart from the reactive grouping of the hydroxamic acid, one more functional group (—OH) in a position favourable for chelation. Naturally, therefore, its presence may change to some extent the nature, stability and the colour of the complexes with respect to those of benzohydroxamic acid.

The maroon cobalt and the light-grey nickel compounds prepared from neutral solutions are of the type[30] $M(C_7H_6O_3N)_2$. Copper[25] and cadmium,[42] on the other hand, give in a faintly ammoniacal medium, green and white 1:1 compounds, respectively. Molybdenum, titanium, niobium, tantalum, zirconium and uranium form oxy compounds, like those with benzohydroxamic acid, and with similar properties.[25]

Vanadate and molybdate yield,[43] at about pH 3, blue-black and yellow solids with a metal to reagent ratio of 1:2. In dilute solutions, at the same pH, composition studies by Job's method show the ratio to be 1:1 but at a higher pH, the latter changes to 1:2, particularly for molybdenum. Iron(III) forms three differently coloured complexes in solution according to the pH. The brown product that separates at pH 1·8–7·0 is soluble in acid, alkali and in organic solvents. Iron combines in a ratio of 1:2 with the acid both in the solid product and in solution.[43] However, an anhydrous iron(III) compound, $Fe(C_7H_6O_3N)_3$, having properties similar to the anhydrous iron(III) benzohydroxamate has also been prepared.[25]

For their gravimetric determinations,[39,42] cadmium is weighed as $Cd(C_7H_5O_3N)$, uranium as $UO_2(C_7H_6O_3N)_2 \cdot 2H_2O$, titanium as $TiO(C_7H_6O_3N)_2$ and zirconium as $ZrO(C_7H_6O_3N)_2$. But in the presence of hydrogen peroxide, titanium[39] and niobium[44] do not react and hence by precipitation at pH 2·5–6·0 in the presence of hydrogen peroxide zirconium is separated from them. The precipitation reaction of this acid towards niobium and tantalum is not strictly quantitative in the presence of citrate, tartrate or oxalate.

Uranium, vanadium, molybdenum, iron, manganese and titanium are all determined photometrically.[25,43,45,46] Uranium, once precipitated, dissolves as the pH of the solution is raised to 7·8–9·2, yielding a solution with increasing ultra-violet absorption.[43] The sensitivity of this colour reaction is 0·1 $\mu g\ cm^{-2}$. Manganese, in alkaline media, develops a greenish-brown colour,[25] with a sensitivity 0·025 $\mu g\ cm^{-2}$, and not the characteristic purple colour.

Coloured complexes of vanadate, molybdate and iron(III) are formed at the respective pH ranges 3·0–3·5, 6·6–7·2 and 8–11 with absorption maxima at 470–480, 400 and 450 nm. The respective sensitivities are 0·017, 0·015 and 0·0125 μg cm^{-2}. The vanadium complex is easily extracted in solvents like ether, ethyl acetate, and amyl and butyl alcohols. With ethyl acetate as the extractant, the vanadium complex is estimated from the complexes of uranium and molybdenum.[43] The colour intensity of the ethyl acetate extract is unaffected by pH changes from 0·8 to 3·5. Titanium gives an intense yellow colour in 6–11 N sulphuric acid. The sensitivity is 0·01 μg cm^{-2} when measured[45] at 390 nm. Extraction of the complex into acetylacetone also is a basis for the determination of titanium.[46]

Salicylhydroxamic acid is a suitable indicator in the chelatometric titration[25] of iron(III) at pH 3.

OTHER HYDROXAMIC ACIDS

Oxalohydroxamic acid, with hydrogen bonded keto and enol forms,[47] has been in use for the direct and indirect colorimetric and indirect volumetric determinations of the reacting elements. For instance, the purple 1:3 iron(III):acid complex[28] and the orange uranium complex, having absorption maxima, respectively, at 500 and 420 nm, are the basis for the direct colorimetric measurements of the ions involved. As little as 0·2 μg of uranium can be detected;[48] the limiting dilution is $1:2\cdot 4 \times 10^5$. The principles of the indirect colorimetric methods for zirconium,[49] calcium[28] and thorium[29] and the volumetric methods for thorium and zirconium have already been described (p. 16).

Cinnamylhydroxamic acid[25,50] reacts somewhat similarly with metal ions to benzo- and salicyl-hydroxamic acids. But towards niobium and tantalum, it reacts differently from oxalate or tartrate solutions from those containing 5% sulphuric acid to those at pH 6·5, both the elements precipitate quantitatively. Niobium forms an insoluble directly weighable oxy-compound, $NbO(C_9H_8O_2N)_3$, stable up to 225° but tantalum gives a compound of indefinite composition.[51]

Phenylacetylhydroxamic acid acts like benzohydroxamic acid at pH 4·5–6·2 for the quantitative separation of tantalum and niobium from each other.[38] It effects complete separation of titanium and zirconium from niobium,[52] which remains in solution at pH 6·5–7·5.

Anthranilohydroxamic acid has been suggested for the spectrophotometric determination of both iron and manganese in the same sample.[53] The orange-red iron(III) hydroxamate, λ_{max} 450 nm, is extracted in butanol at pH 4·5 and manganese in the aqueous phase is determined as its wine-red complex, formed above pH 9·2 (λ_{max} 490–500 nm).

The solubility of the metal hydroxamates, derived from the heterocyclic hydroxamic acids, decreases in the order: isonicotino, nicotino, picolino, quinaldino. In the very same order, their extractability in organic solvents increases. In their reaction towards manganese(II) at or above pH 9, while the nicotino and isonicotino derivatives, give stable and sensitive red-violet colours, the picolino derivative imparts a faint colour which soon turns turbid and the quinaldino compound gives a pale yellow precipitate.[54] Thus only nicotino- and isonicotino-hydroxamic acids are suitable for the determination of manganese by their colour reaction, with sensitivities at 480 nm of 0·014 and 0·013 $\mu g\ cm^{-2}$. Studies indicated the presence of a 1:3 manganese:acid complex in solution.[55] With iron(III) and vanadium(V), a series of colours are developed as the pH values of the solutions are varied. As expected, iron forms three complexes, 1:1, 1:2 and 1:3 (metal:acid) which are violet, red and golden yellow, respectively, in solution. The variety of coloured products given by vanadium on reaction in aqueous and in 50% ethanolic solutions [56,57] are all 1:3. Molybdenum(VI) gives in solution a 1:2 yellow complex. Methods for the spectrophotometric determinations of iron, vanadium and molybdenum have been worked out.

Quinaldinohydroxamic acid yields with iron a red colour at pH 1·2. As the pH is raised to 1·5 a red precipitate is formed which dissolves to give a golden-yellow solution at a pH above 10. The red precipitate can be extracted in isobutanol from aqueous solutions of pH 3–9. The dirty purple precipitate of vanadium gives an intense orange-red solution on extraction with isobutanol from solutions at pH 3·0–4·7. Iron and vanadium are determined in the extracts[54] by measurement at 450 nm.

While benzo- and phenylacetyl-hydroxamic acids, in presence of tartrate or citrate, fail to precipitate niobium and tantalum, quinaldinohydroxamic acid, at pH 5–9 (ammoniacal), produces, in the presence of tartrate, citrate, oxalate and EDTA, yellow precipitates with both the elements. The niobium compound, $NbO(C_{10}H_7O_2N_2)_3$, which is soluble in chloroform, acetone and ethanol, is stable up to 250° and so can be weighed directly. However, the tantalum precipitate, which is moderately soluble in ethanol and slightly so in chloroform, needs ignition to the pentoxide as the final weighing form.[58]

2-Naphthohydroxamic acid in methanol combines with vanadate in a 2:1 ratio to impart an intense red-orange colour.[59] This has been used in a method for the determination of vanadate that has a sensitivity of 0·009 $\mu g\ cm^{-2}$.

The violet reaction products of vanadate with thiophene-2-hydroxamic acid are always 1:3 complexes, irrespective of whether the solution is maintained at pH 3·5 (λ_{max} 590 nm) or 7·7 (λ_{max} 550 nm). The extractability of the product formed in acidic solution has been used to devise a method for the determination of vanadium.[60]

Hydroxylamines

The instability and lack of selectivity of cupferron led to the search for a suitable substitute. Though this attempt has not been very successful, several N-substituted phenylhydroxylamines:

$$C_6H_5-\underset{\underset{\text{R}}{|}}{N}-OH$$

where R=CHO, $COCH_3$, COC_6H_5(I), $CONH_2$, $CONHC_6H_5$, $CSNH_2$, $CSNHC_6H_5$, $CH:NC_6H_5$, $CSNHCH_2CH:CH_2$ and $N:NC_6H_5$, have been studied.[61] Of these, N-benzoylphenylhydroxylamine(I) appears to be the most interesting and promising. It already finds extensive analytical applications. Many other N-acyl substituted phenylhydroxylamines and an N-benzoyl derivative of naphthylhydroxylamine have since been prepared and their reactivities investigated. These compounds are usually soluble in organic solvents, but are slightly so in water. They are white solids, with the exception of 3,5-dinitrobenzoylphenylhydroxylamine which is yellow. They melt with or without decomposition and are stable to heat, light, air and dilute hydrochloric or sulphuric acid. They can be stored dry for years. However, dilute nitric acid, ammonia or alkali, and solutions of permanganate or dichromate decompose the compounds.

A few aromatic hydroxylamines of various acidities have been examined for their reaction towards copper and tin.[62] The pH of 1% solutions of these hydroxylamines in ethanol are in the order N-benzoylphenyl > o-ethoxybenzoylphenyl $\simeq$ o-iodobenzoylphenyl > naphthoylphenyl > 2,4-dichlorobenzoylphenyl > furoylphenyl > 3,5-dinitrobenzoylphenyl > benzoylnaphthyl. As the acid character of these compounds increases, the pH required for the complete precipitation of copper increases. By the shifting of the electron density towards the electron attracting groups the molecule is rendered more acidic and the coordinating ability of the oxygen atom of the carbonyl group is more reduced. Thus by the enhancement of acidity of the hydroxylamines, their copper complexes become less stable. Such an effect is more marked with tin. The benzoylphenyl derivative precipitates tin completely from 8% hydrochloric acid, whereas the naphthoylphenyl, benzoylnaphthyl and 2,4-dichlorobenzoylphenyl need a much lower hydrochloric acid concentration (about 1%) for the same purpose. Furoylphenyl, 3,5-dinitrobenzoylphenyl, o-ethoxybenzoylphenyl and o-iodobenzoylphenyl derivatives, under the same condition, precipitate tin incompletely. The latter two may be suggested to exert some steric effect as well. From a comparison of the reactions of N-naphthoylphenylhydroxylamine and N-benzoylnaphthylhydroxylamine with copper, it appears that there is no marked difference in the effect of substituents on the phenyl ring whether attached to the carbonyl or to the oxime group.

When the reactions of the aromatic phenylhydroxylamines are compared with those of the aliphatic derivatives, very interesting and important differences are observed.[63] In solutions containing more than 1% of hydrochloric acid, tin, titanium, zirconium and vanadium are precipitated by the benzoyl, furoyl and thenoyl derivatives, whereas the hexanoyl, heptanoyl and cyclohexanoyl derivatives give somewhat cloudy products which cannot be filtered. The nicotinyl derivative, however, forms a soluble hydrochloride. In presence of sodium acetate, iron combines with benzoyl-, thenoyl-, furoyl- and nicotinyl-phenylhydroxylamines to produce brown 1:3 metal:acid complexes soluble in organic solvents. Aliphatic derivatives such as hexanoyl-, heptanoyl- and cyclohexanoyl-phenylhydroxylamines, on the other hand, yield an oily product or a fine precipitate. The molar extinction coefficient of the iron chelates depend on the extent of the π-electron system and hence on the nature of the attached group (aromatic or aliphatic). For those with aromatic linkages, the extinction values are naturally higher compared to others with aliphatic groups. However, of the aromatic derivatives, thenoyl- and furoyl-phenylhydroxylamines give colour reaction of greatest sensitivity. Copper precipitation is pH dependent, but even in this instance the aliphatic and aromatic derivatives show distinct differences. The lowest pH for the complete precipitation of copper is 4·5 with aliphatic phenylhydroxylamines, whereas it is only 3 when the aromatic phenylhydroxylamines are the precipitants. The ability to precipitate metal complexes, therefore, rests on the nature of the ring attached to the carbonyl group; but the complex stability in ethanol of the 1:3 iron chelates decreases with the increase of the acidity of the organic molecules, irrespective of the nature of the ring attached to the carbonyl group.

A critical study on the use of benzoyl-o-tolyl- (I), benzoyl-m-tolyl- (II), benzoyl-p-tolyl- (III), benzoyl-p-chlorophenyl- (IV), benzoylphenyl- (V) and phenylacetylphenyl- (VI) hydroxylamines as spectrophotometric reagents[64-66] for vanadate shows that the violet or reddish-violet species formed in the presence of hydrochloric acid and the yellow product formed at pH 4·8–6·0 in the presence of ethanol are both extractable into chloroform. The maximum absorption regions of these extracts are at 510–530 nm for the former and 440 nm for the latter. On the addition of ethanol in increasing quantities to the former, the absorption maximum shifts gradually from 510–530 to 440 nm. This shift suggests the formation of different complexes, which are of the types $[VOR_2Cl]$ and $V_2O_3R_4$ (RH is the hydroxylamine derivative) as proved by the isolation of these complexes in the solid state.[67] The pK_a values of these hydroxylamines as determined in an ethanol–water mixture show the acid dissociation constants to be in the order IV > V > I ≃ II ≃ III > VI.

The reaction sensitivity of these hydroxylamine derivatives towards vanadate increases with their basicity, except for phenylacetylphenylhydroxylamine, which though strongly basic is the least sensitive. This distinctive behaviour of the latter is due to its having a methylene group between the phenyl ring and the carbonyl group, whereas in the others substitution is on the phenyl ring attached to the oximino group. The intervening methylene group is responsible for the lower acid dissociation and sensitivity values. Moreover, all these hydroxylamines behave alike towards interfering ions when the pH of the medium is 4·8–6·0. In hydrochloric acid solution, their reactions towards one or two ions are somewhat different. Such differences cannot be explained from the electronic effects of the substituents. For instance, N-benzoylphenylhydroxylamine (pK_a = 9·8) is less selective than N-benzoyl-p-chlorophenylhydroxylamine (pK_a = 9·6) which is again less selective than N-benzoyl-p-tolylhydroxylamine (pK_a = 9·92). Steric effects only can explain this. The methyl group at the 2-position imparts steric strain and thus reduces the stability of the complexes. In other words, it increases the selectivity of the vanadium reaction, as is found with the o-tolyl derivative, which behaves as a specific reagent for vanadium. Reagents (II) and (III) show similar behaviour, perhaps because of the lack of basic geometry required for the overlap of the orbitals of the metals and the ligands. The same can be said of (VI) in which the phenyl ring attached to methylene group has free rotation. As the group responsible for such steric strain moves further away from the reactive atomic grouping less steric effect is experienced and hence the molecule

N–benzoyl-o-tolylhydroxylamine

becomes less selective in its reaction. Thus the selectivity of their reactions are as I > II > III > VI > IV > V. Again with the decrease in the bulkiness of the group $CH_3 \to Cl \to H$ at the remote para position there is an orderly decrease in selectivity as III > IV > V.

In a recent study[68] of the coloured chloroform extracts of the vanadium(V) reaction products with N-arylhydroxylamines:

$$\begin{array}{c} R_1 - C = O \\ | \\ R_2 - N - OH \end{array}$$

it is noticed that the substitution of the phenyl group (R_1) attached to the carbon atom by an aliphatic chain derived from n-butyric, lauric or palmitic acid causes a slight hypochromic and hypsochromic effect, and the colour changes to reddish-violet. The length of the saturated aliphatic chain, however, has little influence on the colour with regard either to the position or the intensity of the absorption band. The hypochromic effect is more prominent with N-phenoxyacetyl- and N-p-chlorophenoxyacetyl-phenylhydroxylamines. Similar replacement of the phenyl group attached to the carbon or the nitrogen atom by a p-tolyl group results in a slight but definite batho- and hyperchromic effect. Such an effect is more pronounced with p-methoxy-benzoyl derivatives.

N-2-Thiophenecarbonyl-p-tolylhydroxylamine and N-2-thiophenecarbonylphenylhydroxylamine have also been proposed as spectrophotometric reagents for the determination of vanadium by extraction into chloroform from hydrochloric acid.[69]

N-Cinnamoylphenylhydroxylamine has been utilized first as a gravimetric reagent for the determination of niobium and tantalum[51] and subsequently as a spectrophotometric reagent[70-72] for the determinations of titanium(IV), niobium(V), vanadium(V), iron(III) and uranium(VI), the last three by the successive extraction of their complexes into solvents such as chloroform and iso-amyl alcohol.

N-Salicylphenylhydroxylamine[73] which is regarded as having two replaceable hydrogen atoms gives with copper(II) almost the same yellowish-green colour as shown by other N-acyl substituted phenylhydroxylamines with no phenolic OH group. Moreover, a suspension of the copper compound in acetone when treated with neutral iron(III) chloride solution, becomes reddish-violet, confirming that in the copper compound the phenolic OH is not taking part in complex formation. By continuous variation and slope ratio methods, iron(III) is shown[74] to produce coloured complexes in 1:1 aqueous ethanol according to the pH of the solutions. A 1:1 violet complex at pH 0·4 and a 1:2 orange-red complex at pH 3·5 are formed. At a higher pH a 1:3 solid product can be isolated.

N-Acetylsalicylphenylhydroxylamine,[75] on the other hand, finds application as a gravimetric and spectrophotometric reagent for the detection and determination of titanium.

THIOBENZOYLPHENYLHYDROXYLAMINE

N-Benzoylphenylhydroxylamine is a very weak acid, due perhaps to hydrogen bonding,[76] and so when the carbonyl oxygen in benzoylphenylhydroxylamine is substituted by even less electronegative sulphur, the acidity of the product, thiobenzoylphenylhydroxylamine, increases greatly. Even though the latter compound is more acidic its metal chelates are more stable than those due to the former, excepting those of manganese where the reverse is true. Such a distinct behaviour on the part of manganese is ascribed to its class (a) acceptor property. As an analytical reagent, the thioanalogue has immense possibilities. Since it is more acidic, it is expected to form a soluble ammonium salt. Its metal complexes are extractable into chloroform and having both sulphur and oxygen as donor atoms, the reagent combines with a wider range of metals than the oxy compound.[77]

The preparation, properties and the analytical applications of N-benzoylphenylhydroxylamine and its analogues are described in subsequent chapters.

References

1. BAUDISCH, O., *Chem. Ztg.* **33**, 1298 (1909); *C.A.* **4**, 557 (1910).
2. HEYN, A. H. A. and DAVE, B. G., *Talanta* **13**, 27 (1966).
3. HILLEBRAND, W. F., LUNDELL, G. E. F., BRIGHT, H. A. and HOFFMAN, J. I., *Applied Inorganic Analysis*, 2nd ed., John Wiley, New York, 1953, p. 116.

4. MAJUMDAR, A. K. and RAY CHOWDHURY, J. B., *Anal. Chim. Acta* **19**, 18 (1958).
5. MAJUMDAR, A. K. and RAY CHOWDHURY, J. B., *Anal. Chim. Acta* **15**, 105 (1956).
6. MAJUMDAR, A. K. and RAY CHOWDHURY, J. B., *Anal. Chim. Acta* **19**, 576 (1958).
7. CHENG, K. L., *Anal. Chem.* **30**, 1941 (1958).
8. STARY, J. and SMIZANSKA, J., *Anal. Chim. Acta* **29**, 545 (1963).
9. FURMAN, N. H., MASON, W. B. and PEKOLA, J. S., *Anal. Chem.* **21**, 1325 (1949).
10. METZ, C. F., *Anal. Chem.* **29**, 1748 (1957).
11. FINKELSHTEIN, D. N. and ELENEVICH, L. P., *Zavod. Lab.* **7**, 665 (1938); *C.A.* **33**, 80 (1939).
12. ELVING, P. J. and OLSON, E. C., *Anal. Chem.* **27**, 1817 (1955).
13. BAUDISCH, O. and HOLMES, S., *Z. Anal. Chem.* **119**, 16 (1940).
14. OESPER, R. E. and FULMER, R. E., *Anal. Chem.* **25**, 908 (1953).
15. ELVING, P. J. and OLSON, E. C., *J. Am. Chem. Soc.* **78**, 4206 (1956).
16. PUROHIT, D. N., *Talanta* **14**, 353 (1967).
17. SOGANI, N. C. and BHATTACHARYYA, S. C., *Anal. Chem.* **28**, 81 (1956).
18. SOGANI, N. C. and BHATTACHARYYA, S. C., *Anal. Chem.* **28**, 1616 (1956).
19. MAJUMDAR, A. K. and SAHA, S. C., *Anal. Chim. Acta* **40**, 299 (1968).
20. MAJUMDAR, A. K. and SAHA, S. C., *Anal. Chim. Acta* **44**, 85 (1969).
21. GUPTA, H. K. L., JAIN, T. C. and SOGANI, N. C., *J. Indian Chem. Soc.* **37**, 531 (1960).
22. GUPTA, H. K. L. and SOGANI, N. C., *J. Indian Chem. Soc.* **38**, 771 (1961).
23. BRANDT, W. W., *Rec. Chem. Prog.* **21**, 159 (1960).
24. BASS, V. C. and YOE, J. H., *Talanta* **13**, 735 (1966).
25. CHAKRABURTTY, A. K., *Proc. Symposium on the Chemistry of Coordination Compounds*, Agra (India), Part III, 1959, p. 235.
26. MILLER, D. O. and YOE, J. H., *Talanta* **7**, 107 (1960).
27. DAVIDSON, D., *J. Chem. Educ.* **17**, 81 (1940).
28. DHAR, S. K. and DAS GUPTA, A. K., *J. Sci. Ind. Research (India)* **11B**, 520 (1952).
29. DHAR, S. K. and DAS GUPTA, A. K., *J. Sci. Ind. Research (India)* **12B**, 518 (1953).
30. MUSANTE, C., *Gazz. Chim. Ital.* **78**, 536 (1948); *C.A.* **43**, 2116 (1949).
31. WERNER, A., *Ber.* **41**, 1062 (1908).
32. PLAPINGER, R. E., *J. Org. Chem.* **24**, 802 (1959).
33. MATHIS, F., *Compt. Rend.* **232**, 505 (1951).
34. STEINBERG, G. M. and SWIDLER, R., *J. Org. Chem.* **30**, 2362 (1965).
35. DAS GUPTA, A. K. and SINGH, M. M., *J. Sci. Ind. Research (India)* **11B**, 268 (1952).
36. WISE, W. M. and BRANDT, W. W., *Anal. Chem.* **27**, 1392 (1955).
37. MELOAN, C. E., HOLKEBOER, P. and BRANDT, W. W., *Anal. Chem.* **32**, 791 (1960).
38. MAJUMDAR, A. K. and PAL, B. K., *Anal. Chim. Acta* **27**, 356 (1962).
39. CHAKRABURTTY, A. K., *Proc. Indian Sci. Congress*, 45th Session, 1958, p. 149.

40. SEABORG, G. T., KATZ, J. J. and MANNING, W. M., *Transuranium Elements* (N.N.E.S.IV-14B), McGraw-Hill Book Co., New York, 1949, p. 1359.
41. MILNER, G. W. C., *Analyst* **81**, 367 (1956); *C.A.* **50**, 12749 (1956).
42. BHADURI, A. S., *Z. Anal. Chem.* **151**, 109 (1956).
43. BHADURI, A. S. and RAY, P., *Z. Anal. Chem.* **154**, 103 (1957).
44. MAJUMDAR, A. K. and MUKHERJEE, A. K., *Anal. Chim. Acta* **22**, 25 (1960).
45. XAVIER, J., CHAKRABURTTY, A. K. and RAY, P., *Science and Culture (India)* **20**, 146 (1954).
46. ALIMARIN, I. P., BORZENKOVA, N. P. and ZAKARINA, N. A., *Zavod. Lab.* **28**, 958 (1961); *C.A.* **56**, 4097 (1962).
47. MONNIER, D. and JEGGE, C., *Helv. Chim. Acta* **40**, 513 (1957).
48. DAS GUPTA, A. K. and GUPTA, J., *J. Sci. Ind. Research (India)* **9B**, 237 (1950).
49. DHAR, S. K. and DAS GUPTA, A. K., *J. Sci. Ind. Research (India)* **11B**, 500 (1952).
50. CHAKRABURTTY, A. K., Private communication.
51. MAJUMDAR, A. K. and MUKHERJEE, A. K., *Anal. Chim. Acta* **22**, 514 (1960).
52. MAJUMDAR, A. K. and PAL, B. K., *Anal. Chim. Acta* **29**, 168 (1963).
53. DUTTA, R. L., *J. Indian Chem. Soc.* **37**, 167 (1960).
54. DUTTA, R. L., *J. Indian Chem. Soc.* **36**, 339 (1959).
55. DUTTA, R. L., *J. Indian Chem. Soc.* **34**, 311 (1957).
56. DUTTA, R. L., *J. Indian Chem. Soc.* **35**, 243 (1958).
57. DUTTA, R. L., *J. Indian Chem. Soc.* **36**, 285 (1959).
58. MAJUMDAR, A. K. and PAL, B. K., *Z. Anal. Chem.* **184**, 115 (1961).
59. BASS, V. C. and YOE, J. H., *Anal. Chim. Acta* **35**, 337 (1966).
60. MINCZEWSKI, J. and TRYBULA, Z. S., *Talanta* **10**, 1063 (1963).
61. SHOME, S. C., *Current Sci. (India)* **13**, 257 (1944).
62. LUTWICK, G. D. and RYAN, D. E., *Can. J. Chem.* **32**, 949 (1954).
63. ARMOUR, C. A. and RYAN, D. E., *Can. J. Chem.* **35**, 1454 (1957).
64. MAJUMDAR, A. K. and DAS, GAYATRI, *Anal. Chim. Acta* **31**, 147 (1964).
65. MAJUMDAR, A. K. and DAS, GAYATRI, *J. Indian Chem. Soc.* **42**, 189 (1965).
66. MAJUMDAR, A. K. and DAS, GAYATRI, *Anal. Chim. Acta* **36**, 454 (1966).
67. MAJUMDAR, A. K., BHATTACHARYYA, B. C. and DAS, GAYATRI, *J. Indian Chem. Soc.* **45**, 964 (1968).
68. TANDON, U. and TANDON, S. G., *J. Indian Chem. Soc.* **46**, 983 (1969).
69. TANDON, S. G. and BHATTACHARYYA, S. C., *Anal. Chem.* **33**, 1267 (1961).
70. DUTT, N. K. and SESHADRI, T., *Indian J. Chem.* **6**, 741 (1968).
71. PRIYADARSHINI, U. and TANDON, S. G., *Analyst* **86**, 544 (1961).
72. ZHAROVSKII, F. G. and SUKHOMLIN, R. I., *Zhur. Anal. Khim.* **21**, 59 (1966).
73. GHOSH, N. N. and BHATTACHARYYA, A., *J. Indian Chem. Soc.* **41**, 311 (1964).
74. GHOSH, N. N. and BHATTACHARYYA, A., *J. Indian Chem. Soc.* **45**, 1103 (1968).
75. SAVARIAR, C. P. and JOSEPH, J., *Anal. Chim. Acta* **47**, 347 (1969).
76. HADZI, D. and PREVORSEK, D., *Spectrochim. Acta* **10**, 38 (1957).
77. BRYDON, G. A. and RYAN, D. E., *Anal. Chim. Acta* **35**, 190 (1966).

CHAPTER 3

PREPARATION AND PROPERTIES OF N-BENZOYLPHENYLHYDROXYLAMINE AND ITS ANALOGUES

N-Benzoylphenylhydroxylamine (BPHA)

$$\text{C}_6\text{H}_5-\underset{\underset{\text{C}_6\text{H}_5-\text{N}-\text{OH}}{|}}{\text{C}}=\text{O}$$

has been prepared according to the method of Bamberger[1] with slight modification.[2]

Procedure

Dissolve phenylhydroxylamine (30 g) in warm water (1200 ml), filter, cool and add to it a small quantity of sodium hydrogen carbonate. Add dropwise with vigorous stirring benzoyl chloride (45 g) and also add periodically in small quantities sodium hydrogen carbonate (30 g) to keep the soln. slightly alkaline during the entire process. Continue stirring for 90 min. Filter off the resulting solid consisting of a mixture of mono- and di-benzoylphenylhydroxylamines, wash with water, triturate the product in a porcelain mortar with a 10% sodium hydrogen carbonate soln., filter and wash with water to remove entrapped benzoyl chloride.

Treat the white product with aqueous ammonia (sp. gr. 0·88) to dissolve the mono-derivative, filter and add the soln. slowly to an excess of (1 + 5) sulphuric acid, well cooled in a bath of ice and salt. Filter under suction the N-benzoylphenylhydroxylamine that separates, wash with water and recrystallize from ethanol. The final product may also be recrystallized from hot water,[3] benzene[4] or acetic acid[5].

Notes

1. The use of 3 N hydrochloric acid, instead of dilute sulphuric acid, has been recommended[3] for the neutralization of the ammonia soln. of the monobenzoyl derivative. This is to avoid the possibility of contamination of the final product with a reddish-brown oil formed as a result of oxidation of the organic substance during neutralization with sulphuric acid.

2. For recrystallization from hot water, dissolve about 0·6 g of BPHA per 100 ml of boiling water, filter hot and cool in an ice bath. For recrystallization from acetic acid, dissolve BPHA in conc. acetic acid and dilute with water.

BPHA forms colourless, needle-shaped crystals, with melting point 121–2°. It is very slightly soluble in cold water. Its solubility in water at 25 ± 0·3°C has been found[4] to be 0·00195 M, a value that compares reasonably well with that of 0·04 g per 100 ml (0·0019 M) obtained earlier.[3] It is soluble in hot water to the extent of about 0·5 g in 100 ml. The organic solvents in which BPHA has been found to be soluble are benzene, ethanol, acetone, ether, iso-amyl alcohol, chloroform, carbon tetrachloride, cyclohexanone, etc. Its solubility in acetic acid increases with the increase in concentration of the acid.[5] The solubilities[6] of BPHA in aqueous alcoholic mixtures and in some other organic solvents are as given in Tables 3.1 and 3.2.

TABLE 3.1. SOLUBILITY OF BPHA IN ETHANOL–WATER MIXTURES AT 22 ± 1°

Ethanol, vol. %	10	20	30	40	50
Solubility g l^{-1}	0·610	0·920	1·394	2·732	6·430
Ethanol, vol. %	60	70	80	90	96
Solubility g l^{-1}	11·320	16·625	33·900	67·865	109

BPHA has been found to be stable towards heat, light and air. It does not decompose on prolonged exposure to the atmosphere and can be stored in dry glass vessels for a considerable length of time (about 3 years).[7]

It is also stable in 8 N sulphuric or hydrochloric acid but it gradually decomposes in 5 N or more concentrated nitric acid. BPHA

TABLE 3.2. SOLUBILITY OF BPHA IN ORGANIC SOLVENTS AT $22 \pm 1°$

Solvent	Solubility (M/l)	Solvent	Solubility (M/l)
Carbon tetrachloride	0·026	Benzene	0·150
Xylene	0·034	Ethyl acetate	0·480
Toluene	0·061	Dichloroethane	0·280
Diethyl ether	0·135	Chloroform	0·580

rapidly decomposes in alkaline or ammoniacal solutions.[5] Thus in 10^{-3}, 10^{-2}, 10^{-1} and 1 N solutions of alkali or ammonia, measurable decomposition begins after 72, 24, 2 hr and immediately after preparation, respectively. Acidified permanganate or dichromate solution oxidizes the reagent, but in the presence of hydrogen peroxide it is stable.[5] The oxidation reaction is not stoichiometric and hence cannot be used for quantitative purposes.[7] Inorganic reducing agents have no obvious effect on the reagent solution.

BPHA is a very weak acid; no ammonium salt is formed when gaseous ammonia is passed through its ethereal solution.[2] The acid dissociation constant has been measured[7] by the isobestic point method and the partition method as $pK_a = 7·97 \pm 0·11$ (at an ionic strength of 1·0). By a two-phase titration method,[4] the value obtained is $8·15 \pm 0·01$ (at an ionic strength of 0·1). By simple pH titration in aqueous solution and in a water–ethanol mixture the values have been found[8] to be 8·30 and 9·8, respectively.

In aqueous solution it gives an absorption maximum[4] at 253 nm with a molar absorptivity of 7650 and at the isobestic points, 240 and 280 nm, it obeys the Lambert–Beer law in aqueous solution and thus behaves like cupferron.[7] The partition coefficient of BPHA between chloroform and water has been evaluated[4] as $10^{2·33 \pm 0·01}$. Later, a value of $216·6 \pm 5·0$ which is 4·5 times the value for the benzene–water system ($45·54 \pm 0·23$) has been obtained.[7] The chloroform extractability of BPHA decreases at pH 6 and also in strong hydrochloric acid solution (> 4 N). Like cupferron, it precipitates many metal ions as chelates, formed by the replacement of the hydrogen of the oxime by the metal ion and by coordination through oxygen of the carbonyl group:

$$\begin{array}{c}C_6H_5\text{—}C\text{=}O\\ |\qquad\quad\;\;\searrow\\ C_6H_5\text{—}N\text{—}O\end{array}\!\!M/n\;(n\text{=charge on the metal ion})$$

BPHA precipitates from 3 N hydrochloric acid solution, titanium, zirconium, hafnium, niobium, tantalum, tin, antimony and vanadium and, at pH 3–6, thorium, scandium, rare-earths, uranium, chromium, aluminium, iron, zinc, cadmium, mercury, copper, nickel, gallium, indium and thallium. Qualitative reactions[7] of the metal ions are shown in Table 3.3. The reaction is more selective at pH 5–6 in the presence of EDTA, under which circumstances only tin, iron, aluminium, titanium, beryllium, uranium, niobium, molybdenum and vanadium precipitate.

Many of these metal-BPHA precipitates are soluble in chloroform.[5,7] The coloured precipitates of vanadium, molybdenum, titanium, cerium(IV), uranium, cobalt, nickel, iron and copper impart colour to the chloroform extracts (Tables 3.3 and 3.4). In Table 3.4 is given the limiting pH of precipitation of the metal ions when a 2% ethanolic BPHA solution (2 ml) is used as the precipitant.

Thermal analysis[9] has shown the relative decomposition temperatures of the metal–BPHA chelates to be $Fe^{3+} < Cr^{3+} < Cu^{2+} < Al^{3+} < Cd^{2+} \sim Mn^{2+} \sim Co^{2+} < Zn^{2+} < Ni^{2+}$.

N-Substituted Phenylhydroxylamines

The method of preparation of other N-substituted phenylhydroxylamines such as N-3,5-dinitrobenzoyl-, N-2,4-dichlorobenzoyl-, N-o-iodobenzoyl-, N-o-ethoxybenzoyl-, N-hexanoyl-, N-heptanoyl-, N-cyclohexanoyl-*, N-furoyl-, N-naphthoyl- and N-thenoyl-phenylhydroxylamines comprises the addition of the acid chloride to phenylhydroxylamine in an ether solution containing pyridine[10] and the purification of the product by recrystallization.[11,12]

The acid chlorides are prepared by refluxing the corresponding acid with thionyl chloride and distilling out the acid chloride produced.

*N-Hexahydrobenzoyl-.

TABLE 3.3. QUALITATIVE REACTIONS OF METAL IONS WITH BPHA

Elements	Oxidation state	pH for pptn.	Colour of ppt.	Colour of chloroform extract
Aluminium	III	3	White	Colourless
Antimony	III	1	White	Colourless
Antimony	V	3 N	White	Colourless
Beryllium	II	5	White	Colourless (floats)
Bismuth	III	3	White	Colourless
Cadmium	II	4	White	Colourless (floats)
Cerium	III	6	White	Colourless
Cerium	IV	3 N	Orange	Orange
Chromium	III	4	Dirty yellow	Yellow
Cobalt	II	5	Pink	Pink
Copper	II	3	Yellow-green	Yellow-green
Gallium	III	3	White	Colourless
Germanium	IV	3	White	Colourless
Hafnium	IV	3 N	White	Colourless
Indium	III	3·5	White	Colourless
Iron	II	5	Red	Red
Iron	III	3	Violet	Violet
Lanthanum	III	6	White	Colourless
Lead	II	4	White	Colourless
Manganese	II	6	Yellowish	Yellowish (floats)
Mercury	I	4	Yellow	Yellow-green (floats)
Mercury	II	2	Yellow	Yellow-green (floats)
Molybdenum	VI	3	Yellowish	Yellowish
Neodymium	III	>6	White	Colourless
Nickel	II	5	Greenish	Greenish
Niobium	V	3 N	White	Colourless
Palladium	II	3	Orange	Pink
Praseodymium	III	6	White	Colourless
Samarium	III	7	White	Colourless
Scandium	III	3	White	Colourless
Tantalum	V	3 N	White	Colourless
Thallium	III	4	White	Colourless
Thorium	IV	3	White	Colourless
Tin	II	3 N	White	Colourless
Tin	IV	3 N	White	Colourless
Titanium	IV	3 N	Yellow	Yellow
Tungsten	VI	3	Yellowish	Yellowish
Uranium	IV	>5	Orange	Yellow
Uranium	VI	5	Greenish-yellow	Yellow
Vanadium	V	3 N	Violet-red	Violet
Yttrium	III	6	White	Colourless
Zinc	II	4	White	Colourless (floats)
Zirconium	IV	3 N	White	Colourless

TABLE 3.4. pH of Interaction and Extractability into Chloroform

Elements	pH range of pptn.	Colour of ppt.	Behaviour in chloroform	Colour of chloroform extract
Titanium(IV)	−0·9 – 10·0	Yellow	Extracted	Yellow
Aluminium(III)	3·0 – 8·0	White	Extracted	Colourless
Chromium(III)	−0·6 – 8·0	Dirty yellow	Extracted	Yellow
Iron(III)	−0·6 – 10·0	Red	Extracted	Red
Iron(II)	—	Red	Extracted	Red
Manganese(II)	5·5 – 9·5	Pale yellow	Floats	—
Zinc(II)	4·0 – 9·0	White	Floats	—
Cobalt(II)	5·0 – 9·0	Pink	Extracted	Pink
Nickel(II)	5·0 – 9·0	Yellow-green	Extracted	Yellow-green
Silver(I)	—	White	Floats	—
Mercury(I)	3·0 – 8·0	Yellow	Floats	—
Mercury(II)	2·0 – 9·0	Yellow	Floats	—
Copper(II)	2·0 – 10·0	Yellow-green	Extracted	Green
Cadmium(II)	5·0 – 9·0	White	Floats	—
Lead(II)	4·5 – 9·0	White	Extracted	Colourless
Vanadium(V)	−0·8 – 7·0	Red	Extracted	Red

General procedure for Preparing *N*-disubstituted Hydroxylamines

Dissolve phenylhydroxylamine (0·04 mole) in ether (200 ml) and add a drop of pyridine to the soln. Add by drops the acid chloride (0·04 mole) while the soln. is stirred mechanically. Keep the soln. at pH 7 by adding periodically drops of pyridine. The total amount of pyridine required is about 0·045 mole. Stir for 1 hr, then heat on a water bath to remove ether. Treat with aqueous ammonia (sp. gr. 0·88) to dissolve the mono derivative. Filter and add the filtrate slowly with thorough stirring to an ice-cold 1:5 sulphuric acid or 3 N hydrochloric acid. Filter off the solid product, wash with water and recrystallize from a water–ethanol soln. or from ether with the addition of heptane.

Procedure for *N*-nicotinylphenylhydroxylamine[12]

Add nicotinyl chloride (0·02 mole) in dry ether (10 ml) slowly to a soln. of phenylhydroxylamine (0·02 mole) in dry ether (200 ml) at 0°. Stir the mixture mechanically for about 90 min. Replace lost ether by the addition of more dry ether (50 ml). Evaporate the ether on a water bath and recrystallize the solid nicotinylphenylhydroxylammonium chloride from alcohol. The melting point of the hydrochloride is 193°C.

Dissolve the *N*-nicotinylphenylhydroxylammonium chloride in the minimum of conc. ammonium soln. Neutralize with 3 N hydrochloric acid and extract repeatedly with ether. Cool the ether soln. in a dry-ice bath; crystals of N-nicotinylphenylhydroxylamine separate.

Procedure for N-benzoylnaphthylhydroxylamine[11]

Dissolve α-naphthylhydroxylamine (0·022 mole) in ether (200 ml). Cool in an ice bath. Add pyridine (0·03 mole) and then drops of benzoyl chloride 0·012 mole) whilst stirring the mixture mechanically. Continue stirring for 1 hr after the complete addition of the acid chloride. Neutralize with dilute hydrochloric acid and add heptane till there is no further precipitation. Filter, wash the solid with water, and dissolve the product in ammonia soln. Filter, and neutralize the filtrate with 3 N hydrochloric acid. Filter off the product, wash with water and recrystallize from ethanol.

To prepare α-naphthylhydroxylamine,[13] in a 500-ml flask, dissolve 10 g (0·05 mole) of α-nitronaphthalene in 200 ml of 95% ethanol and 50 ml of ether. Cool to 0° and saturate the soln. with anhydrous ammonia. Pass hydrogen sulphide gas through the soln. in a rapid stream until crystals of ammonium sulphide fill the flask. Keep it well stoppered overnight in the cold, and at least for 24 hr. Pour the contents into a large vol. of ice water, wait for 30 min, filter off the precipitated α-naphthylhydroxylamine, wash several times with water and dry.

The physical properties of some substituted hydroxylamines are given in Tables 3.5 and 3.6.

Whereas BPHA is basic, N-furoylphenylhydroxylamine is more acidic (Table 3.5). Hence, N-furoylphenylhydroxylamine is capable of forming an ammonium salt. All these hydroxylamine derivatives are white crystalline substances, except N-3,5-dinitrobenzoylphenylhydroxylamine, which is yellow. They can be stored for 2 years or more without any decomposition; they melt with decomposition. They are slightly soluble in water but are freely soluble in most organic solvents.

In neutral solutions, with 1 ml of a 1% ethanolic solution of the compounds of Table 3.5, manganese(II), lead and aluminium give white precipitates, uranium(VI), copper and iron(III) yield yellow, greenish-yellow and orange precipitates, respectively. In more than 1% hydrochloric acid, however, only tin(II), tin(IV), titanium, zirconium and vanadate give precipitates, the colour of which is white for tin and zirconium, yellow for titanium and purple for vanadium. Vanadium(V) and zirconium are completely precipitated, from 10% and 5% hydrochloric acid, respectively by all the above-mentioned hydroxylamines.[11]

Chromate and permanganate in acidic solution oxidize the compounds and thus are reduced to chromium(III) and manganese(II).

TABLE 3.5 PHYSICAL PROPERTIES OF SOME N-SUBSTITUTED HYDROXYLAMINES[11]

Substituted hydroxylamines	m.p.	Water solubility (g per 100 ml)	Apparent pH of 1% solution of reagent in ethanol	Lowest pH for the complete pptn. of Cu
N-Benzoylphenyl- C_6H_5—C=O | C_6H_5—N—OH	121°	0·040	7·50	3·0
N-o-Ethoxybenzoylphenyl- $C_6H_4(OC_2H_5)$—C=O | C_6H_5—N—OH	103°	0·011	7·30	3·3
N-o-Iodobenzoylphenyl- C_6H_4I—C=O | C_6H_5—N—OH	128°	0·010	7·30	3·0
N-Naphthoylphenyl- $C_{10}H_7$—C=O | C_6H_5—N—OH	129°	0·003	7·20	3·4
N-3,5-Dinitrobenzoylphenyl- $C_6H_3(NO_2)_2$—C=O | C_6H_5—N—OH	133°	0·007	5·80	4·2
N-Furoylphenyl- C_4H_3O—C=O | C_6H_5—N—OH	134°	0·013	6·10	3·0
N-2,4-Dichlorobenzoylphenyl- $C_6H_3Cl_2$—C=O | C_6H_5—N—OH	137°	0·006	6·80	3·7
N-Benzoylnaphthyl- C_6H_5—C=O | $C_{10}H_7$—N—OH	164°	0·013	4·50	5·5

TABLE 3.6 PHYSICAL PROPERTIES OF SOME N-SUBSTITUTED HYDROXYLAMINES

Substituted hydroxylamines	m.p.	Apparent pH of 1% solution of reagent in ethanol	Lowest pH for the complete pptn. of Cu
N-Thenoylphenyl- C_4H_3—S—C=O $\quad\quad\quad\quad\mid$ C_6H_5—N—OH	97°	6.85	3.0
N-Nicotinylphenyl- C_5H_4N—C=O $\quad\quad\quad\mid$ C_6H_5—N—OH	134°		
N-Cyclohexanoylphenyl-* C_6H_{11}—C=O $\quad\quad\mid$ C_6H_5—N—OH	124°	7.25	4.5
N-Hexanoylphenyl- C_5H_{11}—C=O $\quad\quad\mid$ C_6H_5—N—OH	67°	7.30	4.5
N-Heptanoylphenyl- C_6H_{13}—C=O $\quad\quad\mid$ C_6H_5—N—OH	60°	7.30	4.5

* N-Hexahydrobenzoylphenyl-

The physical properties of the hydroxylamines[12] in Table 3.6 are similar to those in Table 3.5. But there is a marked difference between the aliphatic and aromatic compounds, with respect to the ability of the metal complexes to precipitate. The nicotinyl derivative forms soluble metal complexes owing to the protonation of the pyridine nitrogen. In > 1% hydrochloric acid the thenoyl, furoyl and benzoyl derivatives give good precipitates with tin(II), tin(IV), titanium, zirconium and vanadate which can easily be filtered, but those due to the hexanoyl, heptanoyl and cyclohexanoyl derivatives are cloudy and non-filterable.

All these hydroxylamines, including the N-benzoylphenyl- and N-furoylphenyl derivatives, give a purple colour with iron at pH 2, but as the pH is raised by adding sodium acetate, only the aromatic derivatives (benzoyl, theonoyl, furoyl and nicotinyl) give precipitates. These are reddish-brown, soluble in organic solvents and contain iron and ligand in a ratio of 1:3. The aliphatic phenylhydroxylamines, on the other hand, yield brown suspensions which change on standing to an oily product (hexanoyl and heptanoyl derivatives), or to a fine precipitate (cyclohexanoyl derivative).

Iron(III) forms with BPHA in ethanolic solution a 1:1 chelate at pH 1·7 with maximal light absorption at 530 nm and a 1:3 chelate at pH 4·8–7·0 having maximal absorption at 440 nm. The stability constants of these chelates are, respectively, 10^5 and 10^{14}. The stability constant ($\log K$) and the molar absorptivity (ϵ) of the 1:3 iron chelates formed with different hydroxylamines, as measured in ethanol, are given in Table 3.7.

TABLE 3.7. STABILITY CONSTANTS AND MOLAR EXTINCTION COEFFICIENTS OF SOME 1:3 IRON(III)–N-ACYLPHENYL-HYDROXYLAMINE COMPLEXES

N-acyl group	$\log K$	ϵ
Benzoyl	14·1	4495
Heptanoyl	14·0	3920
Hexanoyl	13·6	3965
Cyclohexanoyl	13·6	4065
Nicotinyl	13·1	4580
Furoyl	12·8	5180
Thenoyl	12·5	5295

With copper(II) at pH 4, aromatic derivatives give light-green precipitates, while the aliphatic derivatives produce bluish-green compounds which precipitate slowly and are contaminated with excess reagent.

N-Acetylphenylhydroxylamine

$$\begin{array}{c} CH_3\!-\!C\!=\!O \\ | \\ C_6H_5\!-\!N\!-\!OH \end{array}$$

An attempt to prepare N-acetylphenylhydroxylamine has been made[14] by the action of acetic anhydride on phenylhydroxylamine.

The method followed[15] for the preparation of the compound from acetyl chloride and phenylhydroxylamine, in presence of a little water, and with the addition of sodium hydrogen carbonate to keep the medium slightly alkaline, is described below.

Procedure

In a 250-ml round-bottomed flask, dissolve phenylhydroxylamine (10·9 g) in benzene (50 ml). Add to it water (5 ml) and sodium hydrogen carbonate (1 g). Dissolve acetyl chloride (8 g) in benzene (10 ml) and add this gradually to the mixture in the flask from a burette. Shake vigorously after each addition till the effervescence ceases. During this reaction, keep the water layer alkaline to litmus by the regular addition of sodium hydrogen carbonate in small amounts (about 10 g required in all). Towards the completion of the reaction, the colour of the mixture changes from yellow to pink as acetyl chloride is added. On shaking vigorously, the colour becomes yellow again. Test a drop of the reaction mixture with Tollen's reagent. If the test gives negative results, the reaction is complete. The time required for complete reaction is 2·5 hr.

Separate the benzene layer and evaporate it on a water bath. Crystallize the solid obtained on cooling, from hot water or from a benzene–petroleum ether mixture to give shining white needle-shaped crystals, m.p. 66·5°.

Note. N-Acetylphenylhydroxylamine is soluble in hot water; the solubility at 33·5° is 4·7 g per 100 g of water. It is slightly soluble in petroleum ether, but freely soluble in ethanol, benzene, ether and sodium hydroxide soln. The solid is very stable when dry and stored in stoppered bottles. If exposed to a moist atmosphere, it decomposes within several months.

N-2-Thiophenecarbonyl-p-tolylhydroxylamine[16]
(N-p-Toly-2-thenohydroxamic acid, p-TTHA)

$$C_4H_3S-C(=O)-N(OH)-C_6H_4CH_3$$

Procedure

Dissolve in cold ether (150 ml), p-tolylhydroxylamine (27 g, 0·22 mole) and keep the soln. at 0° or less, by external cooling. While stirring the soln. mechanically, add dropwise over 1 hr pyridine, (18–20 ml) and 2-thiophenecarbonyl chloride (29·3 g, 0·20 mole), from two dropping funnels. Always keep the mixture basic during the reaction. Generally an orange-yellow granular precipitate or sometimes a sticky brownish material appears. Decant the ether layer, evaporate at room temperature under vacuum and mix any solid material thus obtained with the main product. Wash the product successively with 2 N hydrochloric (2 × 25ml) followed by water (3 × 50 ml).Triturate the product in a mortar with a saturated sodium hydrogen carbonate soln. to remove any acidic material. Filter, wash with water and then treat several times with aqueous ammonia (sp. gr. 0·88) to dissolve the mono-2-thiophenecarbonyl derivative. Filter the ammonia soln. and add this dropwise to ice-cold 6 N hydrochloric acid with stirring. Filter the white crystalline precipitate of N-2-thiophenecarbonyl-p-tolylhydroxylamine, wash with cold water and dry. Crystallize from a mixture of benzene and petroleum ether (m.p. 123°).

To prepare 2-thiophenecarbonyl chloride, reflux 2-thiophenecarboxylic acid (30 g) with thionyl chloride (30–35 ml) on a steam bath and distil the product under vacuum (b.p. 65–70° at 5 mm of mercury).

Note. Both N-2-thiophenecarbonyl-p-tolylhydroxylamine and N-2-thiophenecarbonylphenylhydroxylamine[16] (N-phenyl-2-thenohydroxamic acid, PTHA) are soluble in ethanol, chloroform, benzene and are sparingly soluble in cold water, but more so in hot water. The chloroform solutions, if stored in amber-coloured bottles, are stable for several days. The solid reagents are very stable to the action of heat, light and air.

Spectral Characteristics of *p*-TTHA and PTHA

In 95% ethanol soln., *p*-TTHA shows two absorption maxima, one at 252 nm and the other at 286 nm. The molar absorptivities at these peaks are 10,000 ± 100 and 12,800 ± 100, respectively. An ethanolic PTHA solution also gives two absorption maxima, at 254 and 285 nm, with molar extinction coefficients of 12,900 ± 100 and 15,300 ± 100, respectively.

N-Benzoyl-*o*-tolylhydroxylamine[8,17,18]

$$\begin{array}{c} C_6H_5-C=O \\ | \\ CH_3C_6H_4-N-OH \end{array}$$

Procedure

Mix *o*-nitrotoluene (30 g) with water (20 ml), ethanol (30 ml) and ammonium chloride (2 g) and add zinc dust (30 g) slowly in small portions to the mixture with vigorous stirring. Maintain the temperature of the liquid between 60 and 70° during the addition of zinc dust. Continue stirring for 30 min after the addition of zinc dust. Filter the hot soln. and wash first with ether (20 ml), and then with water.

Cool the filtrate in an ice bath, dilute with water to 200 ml and add a small amount of sodium hydrogen carbonate to make it slightly alkaline. Stir the soln. vigorously while adding benzoyl chloride (15 g) dropwise. During this addition add in small portions, from time to time, sodium hydrogen carbonate to keep the solution slightly alkaline. Continue adding benzoyl chloride until a drop of the reaction mixture gives a negative test with Tollen's reagent. Continue stirring for 1 hr. Filter the resulting solid, wash with water, triturate in a mortar with a 10% sodium hydrogen carbonate soln., filter and wash with water. Extract the solid product with aqueous ammonia (sp. gr. 0·88) to dissolve the monoderivative, filter and add the ammonia soln. to ice-cold (1 + 6) sulphuric acid, when N-benzoyl-o-tolyhydroxylamine separates. Filter, wash with water and purify by recrystallization from aqueous ethanol as white needle-shaped crystals, m.p. 104°.

N-Benzoyl-p-tolylhydroxylamine[8,17,18] and N-benzoyl-m-tolylhydroxylamine[8,17,18]

Procedure

Use p- and m-nitrotoluene to prepare, respectively, N-benzoyl-p- and N-benzoyl-m-tolylhydroxylamines, and follow the same procedure as described for N-benzoyl-o-tolylhydroxylamine.

During the addition of ammonia soln. to the ice-cold (1 + 6) sulphuric acid, the N-benzoyl-m-tolylhydroxylamine that separates is a sticky brown semi-solid. For purification, filter, wash with water and treat this with petroleum ether, when yellowish-white crystals separate. Recrystallize from aqueous alcohol, m.p. 70°. The p-derivative has a m.p. of 108°.

N-Benzoyl-o-, N-benzoyl-p- and N-benzoyl-m-tolylhydroxylamines are all very slightly soluble in cold water but more so in hot water, freely soluble in ethanol, chloroform, benzene, xylene and amyl alcohol.

N-Cinnamoylphenylhydroxylamine[19,20]

$$C_6H_5-CH=CH-C(=O)-N(C_6H_5)-OH$$

Procedure

Add to a cold soln. of phenylhydroxylamine (30 g) in water (1200 ml), with thorough mechanical stirring and within a period of 90 min, a soln.

of cinnamoyl chloride (53 g) in ether (75 ml). During this operation, keep the soln. faintly alkaline by the addition of sodium hydrogen carbonate in small portions from time to time. Continue stirring for another 45 min. Filter the resulting solid, wash with water and then follow the procedure described for N-benzoylphenylhydroxylamine. Recrystallize from ethanol as pale-green crystals, m.p. 158–60°.

The compound is insoluble even in hot water but soluble in most organic solvents.

N-Phenylacetylphenylhydroxylamine[20]

$$C_6H_5-CH_2-C=O$$
$$\quad\quad\quad\quad\quad\quad |$$
$$\quad\quad\quad C_6H_5-N-OH$$

Procedure

To a cold soln. of phenylhydroxylamine (15 g) in water (600 ml), add a small amount of sodium hydrogen carbonate to make the soln. faintly alkaline and then drop in phenylacetyl chloride (25 g). Stir and keep the soln. slightly alkaline by the addition of sodium hydrogen carbonate from time to time during the addition of the acid chloride. Continue stirring for some time and isolate the white crystals of N-phenylacetylphenylhydroxylamine as above. Recrystallize from ethanol. The product melts at 90–91°.

The compound is slightly soluble in cold water but more so in hot water and is soluble in ethanol, chloroform, benzene, xylene, cyclohexanol and amyl alcohol.

N-Benzoyl-p-chlorophenylhydroxylamine[21]

$$C_6H_5-C=O$$
$$\quad\quad\quad\quad |$$
$$p\text{-}ClC_6H_4-N-OH$$

Procedure

Dissolve air-dried p-chlorophenylhydroxylamine (15 g) in hot water (600 ml). Filter, and add benzoyl chloride (10 g) dropwise to the hot filtrate. Stir and keep the soln. slightly alkaline throughout the period of addition of the acid chloride by adding periodically small portions of sodium hydrogen carbonate. Test for completion of the reaction by using

Tollen's reagent. Stir for another 30 min after the completion. Isolate N-benzoyl-p-chlorophenylhydroxylamine by following the method given for N-benzoyl-o-tolylhydroxylamine. Recrystallize from ethanol as needle-shaped crystals, m.p. 155°.

The p-chloro- derivative is not soluble in cold water, is very sparingly soluble in hot water and is completely soluble in ethanol, chloroform and other organic solvents.

The acid dissociation constants of a few of the substituted hydroxylamines as determined[8] by pH titration in 1:1 ethanol–water and in water are given in Table 3.8.

TABLE 3.8. ACID DISSOCIATION CONSTANTS OF SOME HYDROXYLAMINE DERIVATIVES

Substituted hydroxylamines	pK_a values*	
	In 1:1 ethanol–water at 22·5°	In water at 25°
N-Benzoyl-o-tolyl-	9·92	8·40
N-Benzoyl-m-tolyl-	9·92	Insoluble
N-Benzoyl-p-tolyl-	9·92	8·40
N-Phenylacetylphenyl-	10·08	8·60
N-Benzoyl-p-chlorophenyl-	9·60	Insoluble
N-Benzoylphenyl-	9·80	8·30

* All values ± 0·04.

Vanadium(V) forms two types of complex with the hydroxylamine derivatives. One in the presence of conc. hydrochloric acid and the other in presence of ethanol at a pH 4·8–6·0. The chloroform extracts of the former show maximal absorptions at 510–530 nm and those of the latter give an absorption maximum at 440 nm. Under both conditions, however, the complexes have[8,21] a ratio of 1:2 (metal:reagent). On the addition of alcohol the absorption maxima of the former extracts change from 510–530 nm to 440 nm.

The complexes formed in strong hydrochloric acid solution, and at a higher pH have on isolation been found to be diamagnetic, as expected, and are of compositions VOL_2Cl and $V_2O_3L_4$, respectively, where LH represents the aromatic hydroxylamine such as the

N-benzoyl-o-tolyl, N-benzoyl-p-tolyl, N-benzoyl-p-chlorophenyl and N-benzoylphenyl derivatives. The compounds of the first type impart violet colours to solutions in acetone, chloroform, carbon tetrachloride and benzene, but in ethanol and pyridine the colour is orange. The complexes of the second type are deep brown and are soluble in all the above-mentioned solvents to which they give an orange colour. However, the colour changes from orange to violet if hydrochloric acid is added to the alcoholic solution of both the complexes.

N-Salicylphenylhydroxylamine[22, 23]

$$o\text{-HOC}_6\text{H}_4-\underset{\underset{\text{C}_6\text{H}_5-\text{N}-\text{OH}}{|}}{\text{C}}=\text{O}$$

Procedure

From a dropping funnel add salicyl chloride (45 g) dropwise to the slightly alkaline ice-cold soln. of phenylhydroxylamine (30 g) in water (800 ml). Stir the mixture vigorously and keep it faintly alkaline, throughout the operation, by adding sodium hydrogen carbonate at intervals. Continue stirring for 1 hr after the complete addition of the acid chloride. Decant the supernatant soln. Wash the sticky greenish residue several times with a 10% sodium hydrogen carbonate soln. Filter, extract with ether, add aqueous ammonia (cold 1N) and shake to dissolve out the mono derivative. Retain the ammonia soln. Add another portion of aqueous ammonia, shake and combine this ammonia extract with the previous one. Repeat this operation several times with small portions of aqueous ammonia. Add the ammonia soln. very slowly to a slight excess of ice-cold (1 + 6) sulphuric acid with vigorous stirring. Keep the temperature below 10°. Redissolve the resulting sticky solid in cold dilute ammonia, treat with charcoal, filter and acidify with cold dilute sulphuric acid as above. Filter and re-crystallize as needle white crystals, m.p. 106°, from aqueous ethanol containing a little animal charcoal.

The compound is soluble in cold or hot water, soluble in acetone, warm ether and ethanol, and also soluble in benzene, chloroform, xylene and amyl alcohol.

Table 3.9 gives the pH ranges for the complete precipitations of some metal ions, as obtained by pH titration.[23] Some of these metal ions give colour reactions at a lower pH and form precipitates at a higher pH. For instance, iron(III) gives a violet colour at a lower pH, but yields a chocolate-red precipitate at a higher pH. Uranium(VI) gives an orange-red colour at a lower pH and a brown precipitate at a higher pH.

TABLE 3.9. pH RANGE FOR PRECIPITATION OF SOME METAL–N-SALICYLPHENYL-HYDROXYLAMINE CHELATES

Compounds*	Colour	pH range for pptn.
Cu(RH)$_2$	Yellow-green	3·0–3·5
Ni(RH)$_2$	Pale green	4·5–5·0
Fe(RH)$_3$	Chocolate-red	1·7–2·0
VO(OH)(RH)$_2$	Violet-black	1·2–1·5
(UO$_2$)$_2$R(RH)$_2$·2H$_2$O	Brown	4·5–5·0

* (RH)$_2$ = N-Salicylphenylhydroxylamine.

N-Acetylsalicylphenylhydroxylamine[24]

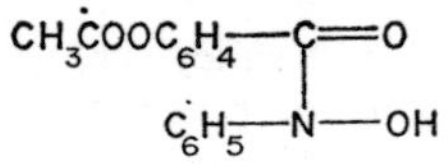

$$CH_3COOC_6H_4-C=O$$
$$C_6H_5-N-OH$$

Procedure

Dissolve phenylhydroxylamine (15 g) in water (200 ml), stir and add to it dropwise acetylsalicyl chloride (7 g), over 5 min, keeping the medium alkaline by the addition of solid sodium hydrogen carbonate. Stir for another 5 min, extract the product with diethyl ether, shake the ethereal layer several times with saturated sodium hydrogen carbonate soln. and then extract the compound into concentrated ammonia soln. Isolate the product by adding the ammonia extract to 4 N hydrochloric acid at 0–4° and crystallize from 1:1 ethanol–water. The product melts at 128°.

N-Benzoylmethylhydroxylamine[25]

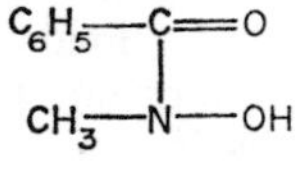

$$C_6H_5-C=O$$
$$CH_3-N-OH$$

Procedure

Add benzoyl chloride (32 g) dropwise with stirring to a well-cooled aqueous soln. of methylhydroxylamine (which has been prepared by the reduction with zinc dust of nitromethane (15 ml) in an ammonium chloride medium). During addition, keep the reaction mixture distinctly alkaline by adding sodium hydrogen carbonate (25 g) in small quantities. Test a drop of the reaction mixture with Tollen's reagent. If the test is negative, stop the addition of benzoyl chloride and stir for another 30 min. Filter under suction. Heat to evaporate the filtrate to dryness under reduced pressure. Extract with methanol and add to the extract small quantities of ether to separate out the inorganic matter. Filter, concentrate the filtrate on a water bath and cool, when the sodium salt of N-benzoylmethylhydroxylamine separates as light-yellow needles.

The salt is very soluble in ethanol and in water; 40·5 g dissolves at 17° in 100 g of water. But it is only sparingly soluble in chloroform, ether and benzene.

N-Substituted Arylhydroxylamines

The method of preparation as proposed[26] for a few more N-acyl substituted arylhydroxylamines such as o-, m- and p-methylbenzoyl-, o-chloro-, p-fluoro- and o-iodo-benzoyl-, o-, m- and p-bromobenzoyl-, p-methoxybenzoyl-, phenylacetyl-, n-butyryl-, lauryl- and palmityl-p-tolylhydroxylamines and also m-methylbenzoyl- and p-fluorobenzoyl-, o- and p-bromobenzoyl-, p-methoxybenzoyl-, phenoxyacetyl-, p-chlorophenoxyacetyl-, n-butyryl-, lauryl- and palmityl-phenylhydroxylamines is given below.

The procedure prescribes the use of the arylhydroxylamine and acid chloride in equimolar proportions for reaction at a low temperature in diethyl ether to dispense with the treatment by concentrated ammonia, as then the N-mono-derivative contains only a negligible amount of di-substituted product, which is removed easily by one or two crystallizations.

Procedure

Take freshly crystallized phenyl- or p-tolyl-hydroxylamine (0·1 mole) dissolved in diethyl ether (150 ml) and a fine suspension of sodium hydrogen carbonate (0·15 to 0·20 mole) in water (25 ml) in a three-necked 500-ml flask provided with stirrer, thermometer and dropping funnel. Cool the mixture to 0° or lower and drop in slowly with stirring over 1 hr freshly distilled acid chloride (0·1 mole) in diethyl ether (100–150 ml). Usually a white precipitate separates. Remove the ether layer and distil off the ether under vacuum. Triturate the light-yellow solid so obtained along with the white precipitate for 15 min with a saturated sodium hydrogen carbonate soln. to remove the acid impurities. Filter the solid product, wash with water, dry in air and crystallize from a mixture of benzene and petroleum ether of boiling range 60–80° or from a mixture of ethanol and water.

The melting points of these N-substituted arylhydroxylamines are given in Table 3.10.

TABLE 3.10. MELTING POINTS OF SOME HYDROXYLAMINE DERIVATIVES

Hydroxylamines	m.p.
N-n-Butyrylphenyl-	81°
N-n-Butyryl-p-tolyl-	63°
N-Laurylphenyl-	80°
N-Lauryl-p-tolyl-	80°
N-Palmitylphenyl-	91°
N-Palmityl-p-tolyl-	85°
N-Phenylacetyl-p-tolyl-	110°
N-Phenoxyacetylphenyl-	154°
N-p-Chlorophenoxyacetylphenyl-	175°
N-p-Fluorobenzoylphenyl-	138°
N-p-Fluorobenzoyl-p-tolyl-	150°
N-o-Chlorobenzoyl-p-tolyl-	139°
N-o-Bromobenzoylphenyl-	113°
N-o-Bromobenzoyl-p-tolyl-	130°
N-m-Bromobenzoyl-p-tolyl-	88°
N-p-Bromobenzoylphenyl-	160–164°d
N-p-Bromobenzoyl-p-tolyl-	180°d
N-o-Iodobenzoyl-p-tolyl-	128°
N-m-Methylbenzoylphenyl-	93°
N-m-Methylbenzoyl-p-tolyl-	112°
N-o-Methylbenzoyl-p-tolyl-	103°
N-p-Methylbenzoyl-p-tolyl-	119°
N-p-Methoxybenzoylphenyl-	130°
N-p-Methoxybenzoyl-p-tolyl-	129°

N-Thiobenzoylphenylhydroxylamine[27] (TBPHA)

$$\begin{array}{c} C_6H_5-C=S \\ C_6H_5-N-OH \end{array}$$

Procedure

Add to phenylhydroxylamine (19·6 g, 0·18 mole) dissolved in the minimal amount of ether, thiobenzoyl chloride (19·2 g, 0·12 mole) quickly with stirring as its purple colour disappears on reaction. Evaporate the yellow soln. under reduced pressure until crystallization begins. Cool in a refrigerator to allow N-thiobenzoylphenylhydroxylamine to crystallize. Recrystallize the yellow solid from aqueous ethanol to give a product of m.p. 102–3°.

Alternatively, for a better yield,[28] extract with conc. ammonia soln. the TBPHA formed in the ether layer, and slowly neutralize with cold 3 N hydrochloric acid, in an ice-salt bath. Purify the precipitated product by recrystallization from ethanol.

The solid is stable indefinitely; as a 1% soln. in ethanol it does not decompose even on storing for several weeks. TBPHA is insoluble in water but is soluble in weakly basic or strongly acidic solutions. It is not stable for more than a few days in basic solutions.

The acid dissociation constants, as calculated from the pH titration curves, suggest TBPHA to be a stronger acid than BPHA; the pK_a values are 10·45 and 8·00 for BPHA and TBPHA, respectively. The stability constants of some 1:2 metal chelates in 50% aqueous dioxan at 25°C are given in Table 3.11. They show the TBPHA chelates to be more stable than the BPHA metal chelates, except those of manganese where the reverse is true.

TABLE 3.11. STABILITY CONSTANTS OF SOME TBPHA AND BPHA CHELATES

Metal ions	Ligand	$\log k_1$	$\log \beta_2$
Manganese(II)	BPHA	5·9	10·8
	TBPHA	5·2	9·7
Nickel(II)	BPHA	7·0	12·5
	TBPHA	8·0	15·5
Copper(II)	BPHA	10·3	18·7
Zinc(II)	BPHA	7·2	13·8
	TBPHA	8·2	15·4

Table 3.12 summarizes the results of comparative qualitative reactions with metal ions as observed by the addition of an ethanolic 1% reagent solution to the metal ions in 1 ml of water or in 1 ml of (1+10) hydrochloric acid. Some of the precipitates impart a colour to the chloroform extract.

50 N-BENZOYLPHENYLHYDROXYLAMINE AND ITS ANALOGUES

TABLE 3.12. REACTIONS OF METAL IONS WITH BPHA AND TBPHA

Metal ions	Colour of precipitate		Colour of chloroform extract		Reaction in (1:10) HCl	
	BPHA	TBPHA	BPHA	TBPHA	BPHA	TBPHA
Silver(I)		White		Pale Yellow	–	+
Aluminium(III)	White	White	Colourless	Colourless	–	–
Gold(III)		Brown		Yellow	–	–
Bismuth(III)		Yellow		Pale yellow	–	+
Cadmium(II)		White		Colourless	–	–
Cerium(IV)	Brown	Cream	Yellow	Colourless	–	–
Cobalt(II)		Brown		Golden	–	–
Copper(II)	Yellow	Brown	Yellow-green	Yellow	–	+
Iron(III)	Violet	Dark brown	Violet	Dark brown	+	+
Mercury(I)	Yellow	Cream	Yellow	Colourless	–	+
Mercury(II)	Yellow	White	Colourless	Unextractable	–	+
Manganese(II)		Yellow-green		Yellow-green	–	–
Molybdenum(VI)		Yellow		Yellow	+	+
Niobium(V)	White	Cream	Colourless	Pale yellow	–	+
Nickel(II)	White	Brown	Colourless	Yellow	–	–
Lead(II)	White	Cream	Colourless	Pale yellow	–	+
Palladium(II)	Grey-green	Buff	Grey-green	Yellow	++	++
Platinum(IV)		Yellow		Yellow	–	–
Tin(IV)	White	White	Colourless	Colourless	+	+ Slow
Tantalum(V)	White	Cream	Unextractable	Colourless	++	++
Thorium(IV)		White		Unextractable	–	–
Titanium(IV)	Yellow	Pale yellow	Yellow	Pale yellow	+	+
Uranium(VI)	Copper	Brown	Copper	Unextractable	–	–
Vanadium(V)	Purple	Green	Purple	Green	++	++
Tungsten(VI)	Cream	Cream	Pale yellow	Pale yellow	–	+
Zinc(II)		White		Colourless	–	–
Zirconium(IV)	White	Yellow	Colourless	Pale yellow	+	+

The precipitation reactions[28] of the metal ions with TBPHA are summarized in Tables 3.13 and 3.14. The results in Table 3.13 are of experiments performed generally with hydrochloric acid. For silver(I), lead(II) and mercury(I), up to 3 M nitric acid is used above which sulphuric acid is employed.

TABLE 3.13. PRECIPITATION REACTIONS OF METAL IONS WITH TBPHA

Metal ion	Detection limit (ppm)	Hydrogen ion concn.			
		0·1	1·0	3·0	6·0
Silver(I)	1	(C)	(C)	(C)	(C)
Bismuth(III)	0·2	(C)	(I)	(—)	—
Cobalt(II)	1	(I)	(—)	—	—
Copper(II)	1	(C)	(C)	(C)	(C)
Iron(III)	5	(C)	(C)	(C)	(C)
Mercury(I)	1	(C)	(C)	(C)	(C)
Mercury(II)	1	(C)	(I)	(—)	—
Molybdenum(VI)	2	(C)	(C)	(C)	(C)
Manganese(II)	0·1	(I)	(—)	—	—
Niobium(V)		(C)	(C)	(C)	—
Nickel(II)	3	(C)	(I)	(—)	—
Lead(II)	0·8	(C)	(—)	—	—
Palladium(II)	1·5	(C)	(C)	(C)	(C)
Platinum(IV)	10	(C)	(I)	(—)	—
Tin(II)	1	(C)	(C)	(I)	(—)
Tin(IV)	1	(C)	(C)	—	(—)
Antimony(III)	10	(C)	(I)	(—)	—
Titanium(IV)	30	(C)	(—)	(—)	(C)
Vanadium(V)	10	(C)	(C)	(C)	(C)
Zirconium(IV)	10	(C)	(C)	(I)	(—)

(C) = complete pptn. (I) = incomplete pptn. (—) = no reaction.

The results recorded in Table 3.14 are visual. Tests requiring disodium EDTA are made with 1 ml of its 0·25 M solution.

TABLE 3.14. PRECIPITATION REACTIONS OF METAL IONS WITH TBPHA

Metal ion	Reaction medium			Stability of ppt. in strong ammonia soln.
	12 M HCl	24 N H$_2$SO$_4$	pH 2–3 plus EDTA	
Silver(I)	(—)	(C)	(C)	(—)
Bismuth(III)	(—)	—	(—)	(C)
Cobalt(II)	—	(—)	(—)	(C)
Copper(II)	(—)	(—)	(C)	(C)
Iron(III)	(—)	(—)	(—)	(C)
Mercury(I)	—	(—)	(C)	(C)
Mercury(II)	(—)	(—)	(C)	(C)
Manganese(II)	—	—	(—)	(—)
Molybdenum(VI)	(C)	(C)	(C)	(—)
Niobium(V)	(C)	(—)	(—)	(—)
Nickel(II)	(—)	—	(—)	(C)
Lead(II)	(—)	—	(—)	(C)
Palladium(II)	(C)	(C)	(C)	(C)
Platinum(IV)	(—)	—	(—)	(—)
Tin(II)	—	—	(—)	(—)
Tin(IV)	(—)	—	(—)	(—)
Antimony(III)	(—)	—	(—)	(—)
Titanium(IV)	(C)	(—)	(—)	(—)
Vanadium(V)	(C)	(—)	(—)	(—)
Zirconium(IV)	(—)		(—)	(—)

Aluminium, cadmium, cerium(IV), thorium, uranium(VI) and zinc precipitate in neutral solutions.

References

1. BAMBERGER, E., *Ber.* **52,** 1116 (1919).
2. SHOME, S. C., *Analyst* **75,** 27 (1950).
3. RYAN, D. E. and LUTWICK, G. D., *Can. J. Chem.* **31,** 9 (1953).
4. DYRSSEN, D., *Acta Chem. Scand.* **10,** 353 (1956).
5. ZHAROVSKII, F. G., *Ukrain. Khim. Zhur.* **25,** 245 (1959).
6. ZHAROVSKII, F. G. and SHPAK, E. A., *Ukrain. Khim. Zhur.* **25,** 800 (1959).
7. YUN-HSIANG, CHIEH, *Candidate's Thesis*, Moscow State University, 1960; *Uspekhi Khim* **31,** 989 (1962).
8. MAJUMDAR, A. K. and DAS, GAYATRI, *Anal. Chim. Acta* **36,** 454 (1966).

9. MEYER, R. A., HAZEL, J. F. and MCNABB, W. M., *Anal. Chim. Acta* **31**, 419 (1964).
10. LUKASCHEWITSCH, W. O., *Ann.* **521**, 198 (1936).
11. LUTWICK, G. D. and RYAN, D. E., *Can. J. Chem.* **32**, 949 (1954).
12. ARMOUR, C. A. and RYAN, D. E., *Can. J. Chem.* **35**, 1454 (1957).
13. WILLSTATTER, R. and KUBLI, H., *Ber.* **41**, 1937 (1908).
14. BAMBERGER, E. and DESTRAZ, H., *Ber.* **35**, 1883 (1902).
15. GUPTA, H. K. L. and SOGANI, N. C., *J. Indian Chem. Soc.* **37**, 769 (1960).
16. TANDON, S. G. and BHATTACHARYYA, S. C., *Anal. Chem.* **33**, 1267 (1961).
17. MAJUMDAR, A. K. and DAS, GAYATRI, *Anal. Chim. Acta* **31**, 147 (1964).
18. DAS, GAYATRI, Ph.D. Thesis, Jadavpur University, 1967.
19. MAJUMDAR, A. K. and MUKHERJEE, A. K., *Anal. Chim. Acta* **22**, 514 (1960).
20. MUKHERJEE, A. K., Ph.D. Thesis, Jadavpur University, 1959.
21. MAJUMDAR, A. K. and DAS, GAYATRI, *J. Indiam Chem. Soc.* **42**, 189 (1965).
22. PAL, B. K., Ph.D. Thesis, Jadavpur University, 1964.
23. GHOSH, N. N. and BHATTACHARYYA, A., *J. Indian Chem. Soc.* **41**, 311 (1964).
24. SAVARIAR, C. P. and JOSEPH, J., Private communication.
25. GUPTA, H. K. L. and SOGANI, N. C., *J. Indian Chem. Soc.* **40**, 15 (1963).
26. PRIYADARSHINI, U. and TANDON, S. G., *J. Chem. Eng. Data* **12**, 143 (1967).
27. BRYDON, G. A. and RYAN, D. E., *Anal. Chim. Acta* **35**, 190 (1966).
28. CASSIDY, R. M. and RYAN, D. E., *Anal. Chim. Acta* **41**, 319 (1968).

CHAPTER 4

GRAVIMETRIC DETERMINATION OF THE ELEMENTS WITH N-BENZOYLPHENYLHYDROXYLAMINE AND ITS ANALOGUES

CUPFERRON and N-benzoylphenylhydroxylamine (BPHA) react under similar condition with many elements, but the metal complexes of the latter reagent, being more stable in mineral acids, offer definite advantage over those of cupferron.

BPHA, as compared to its other analogues, has been the most extensively used as an analytical reagent for the gravimetric determination of a large number of elements and their separation from other ions under conditions controlled mostly by pH. EDTA (disodium salt) and tartaric acid have been used as masking agents, the latter being used mainly to keep the easily hydrolysable metal ions in solution.

The most interesting and important uses of the reagent are the determination and separation from each other of niobium and tantalum, the direct determination of niobium in presence of tantalum and other ions, the separation of niobium, tantalum and titanium from each other, and the determination of zirconium in presence of niobium, tantalum, titanium and vanadium.

Table 4.1 gives an account of the separations of the elements so far achieved, their pH of complete precipitation and their mode of determination. The reactions of thallium, vanadium, chromium, manganese, silver, zinc, cadmium and antimony with BPHA have not been studied.

As precipitates involving BPHA are often directly weighable, homogeneous precipitation has not found much favour in such reactions.

Of the BPHA analogues, mention may be made of N-cinnamoyl-phenylhydroxylamine, N-benzoyl-o-tolylhydroxylamine, N-salicyl- or N-acetylsalicyl-phenylhydroxylamine and N-thiobenzoylphenyl-hydroxylamine which have been used, respectively, for the determination of niobium and tantalum in presence of various other ions, the separation of niobium and tantalum from each other and from other ions, and their determinations, the determination of titanium and that of iron and copper in the presence of different ions. The N-furoyl and N-methyl derivatives require further study as the former has some interesting properties and the latter is very soluble in water.

1. Applications of N-Benzoylphenylhydroxylamine

DETERMINATION OF COPPER, IRON AND ALUMINIUM AND THEIR SEPARATION FROM DIVERSE IONS
(Shome[1])

Copper, iron and aluminium are quantitatively precipitated by BPHA in the pH ranges 3·6–6·0, 3·0–5·5 and 3·6–6·4, respectively. The precipitates, of composition $Cu(C_{13}H_{10}O_2N)_2$, $Fe(C_{13}H_{10}O_2N)_3$, and $Al(C_{13}H_{10}O_2N)_3$, respectively, are green, red and white with melting points as 198–9°, 187–8° and 238–9°. These compounds are soluble in many organic solvents, and are more or less so in ethanol. Moderately concentrated mineral acids decompose the compounds. All the three elements have been determined in presence of other ions, for example, copper in presence of lead, mercury(II), cadmium, cobalt, beryllium, zinc, manganese, nickel and uranium(VI), iron(III) in presence of cobalt, nickel, manganese, uranium(VI) and zinc, and aluminium in presence of beryllium, cobalt, nickel, manganese, uranium(VI) and zinc. Copper has also been determined with separation from phosphate, arsenate and arsenite in presence of tartaric acid.

TABLE 4.1. GRAVIMETRIC DETERMINATIONS WITH BPHA

Elements	pH of pptn.	Mode of determination	Separation from	Refs.
Aluminium	3·6–6·4	D	Be, Co, Ni, Mn, U(VI), Zn	1
Beryllium	5·5–6·5	D	Fe(III), Al, Ti	13
Bismuth	6·0–6·8	D	Fe(III), Ti, W(VI), Mo(VI), V(V), Pb, As(V), Sb(III), Th, U(VI), La, Be, Sn(IV), Al, Cu, Co, Ni, Zn, Mn, Cd, Hg(II), Pd	20
Cerium(III)	6·5–7·5	I	Th	7
Cobalt	5·5–6·5	D	Cu	6
Copper	3·6–6·0	D	Pb, Hg(II), Cd, Co, Be, Zn, Mn, Ni, U(VI), As(III and V), PO$_4$	1
Gallium	2·0–3·0	D	Al, Cu, Th, Ce(III), Ti, U(VI), In, Be, Fe(II), Zn	15, 16
Germanium	≥ 2 N HClO$_4$	I	Fe(III), Cu, Zn, Sn, Ni, Ca	32
Indium	4·3–8·0	D	Co, Ni, Mn, U(VI), Zn	15
Iron(II)	3·0–5·5	D	U(VI), Ce(IV), Fe(II and III), Al, Ga, In, Sc, Zn, Cu, Ni, Th	1
Lanthanum	6·4–7·2	D		18
Magnesium	8·0–9·0	I	Be, Cu, Ni, Co, Zn, Al, Fe(III), Th	33
Mercury(II)	3·0–6·0	D	Zn, Co, Ni, Pb, Ag, Tl(I), Bi, Sb, Cd, Sn(IV), In, As(V), W(VI), Mo(VI)	19
Molybdenum(VI)	0·01–2·5 N HCl	D or I	Cu, Co, Ni, Fe, V(V), Cr(VI)	12
Nickel	5·5–6·5	D	Cu	6
Niobium	< 0·0	I	Zn, Mn, Ni, Co, Cu, Cd, Hg(II), Be, Mg, U(VI), Cr(III and VI), Fe(III), Al, Ce(III and IV), As(III and V), Th, PO$_4$ (steel)	27, 28

GRAVIMETRIC DETERMINATION OF THE ELEMENTS WITH BPHA

Element	Acidity	D or I	Elements that do not interfere	References
Scandium	3·5–6·5	D	Ta, Fe(III), U(VI), Be, Th, Cr(III and VI), Al, Rare-earths, Ce(III and IV), Cu, Cd, Bi, Pb, Hg(II), As(III and V), Sb(III), Sn(IV), Zn, Mn, Ni, Co, Ca, Sr, Ba, Mg, W(VI), PO_4	24, 25, 26, 29, 30, 31
Tantalum	5·4	I	Zr, Rare-earths	5
	<0·0–1·5	I	Ti, Zr, Nb	23
			Zn, Mn, Ni, Co, Cu, Cd, Hg(II), Be, Mg, U(VI), Cr(III and VI), Fe(III), Al, Ce(III and IV), As(III and V), Th, PO_4 (steel)	27, 28
Thorium	4·0–8·5	D or I	La, Ce(III), Rare-earths, Ti, Zr, Ga, In, Fe(III), Al, V(V), U(VI)	7, 8, 9
Tin(IV)	0·1–0·5 N HCl	D	Cu, Pb, Zn	2
Titanium	0·12–0·5 N HCl	I	Mo(VI), Th, Ce(III), V(V), Fe(III), Zr, Cu, Bauxite	21
	0·02–0·5 N H_2SO_4	D	Th, Ce(III), U(VI), Cr(III), Al, Zn, Mn, Ni, Co, Fe(III), V(IV), Cu, PO_4	22
Tungsten(VI)	0·05–1·0 N HCl or 0·5–1·0 N HCl	D	Mo(VI), U(VI), Ti, Fe(III), V(V)	17
Uranium(VI)	5·2–5·6	D	Pb, Bi, Fe(III), Ti, Zr, Mo(VI), Ce(III), Al, Th	14
Zirconium	2·4 N HCl or 3·6 N H_2SO_4	I	Ti, Nb, Ta, Fe(III), Al, Cr, Rare-earths, V(V)	10
	0·5–0·6 N HCl	I	Cu, Al, Bi, Cd, Cr(III), Pb, Mg, Mn, Ni, Zn, U(VI), Th, Yt, Ce(III), Sm, La, Pr, Gd, Nd, Dy	11
	0·5–0·6 N H_2SO_4	D		

Determination of Copper, Iron and Aluminium

Neutralize with dilute ammonia soln. the soln. containing copper (25 mg), iron (15 mg) or aluminium (10 mg) and then add 1·08 N sulphuric acid (5 ml). Dilute to 400 ml and heat to boiling. Add carefully a warm soln. of BPHA, containing at least 1·75 times the theoretical amount of reagent, in ethanol (15–20 ml). Adjust the pH to 4 by adding a 10% sodium acetate soln. (10 ml), heat at 65° on a water bath with occasional stirring for 1 hr for copper precipitation and 2 hr for iron or aluminium precipitation. Filter the precipitate through a No. 4 sintered glass crucible, wash with hot water (at 45° for aluminium) and dry at 110° to constant weight.

Weight of copper complex × 0·1302 = weight of copper.
Weight of iron complex × 0·08064 = weight of iron.
Weight of aluminium complex × 0·04065 = weight of aluminium.

Notes

1. The reagent soln. must not fall on the side of the beaker. The final soln. must not contain more than 5% of ethanol during the precipitation of aluminium, as the aluminium complex is moderately soluble in ethanol.

2. Beryllium, cadmium, cobalt, nickel, manganese, zinc and uranium(VI) remain in solution under these conditions.

Separation of Copper from Lead

Add glacial acetic acid (5 ml) to the neutralized soln., dilute to 400 ml and heat to 65°. Add 1·75 times the theoretical amount of BPHA in ethanol (10 ml), adjust to pH 4 by adding a few ml of 10% sodium acetate soln. and then proceed as suggested for the copper determination.

Separation of Copper from Mercury

Follow the procedure for the separation of copper from lead, but add some chloride ions to keep mercury in solution.

Separation of Copper from Phosphoric, Arsenic and Arsenious Acids

Add to the soln., acidified with sulphuric acid, a 10% soln. of Rochelle salt (5 ml for each 0·1 g of P_2O_5, As_2O_5 or As_2O_3), dilute with water to 400 ml and heat. Add the required amount of BPHA in ethanol (10 ml) followed by a 10% sodium acetate soln. to adjust the pH of the soln. to 4·5–4·7 and then proceed as above. Tartrate in large excess does not affect the precipitation of copper, iron and aluminium.

DETERMINATION OF TIN
(Ryan and Lutwick[2])

BPHA was originally presumed to combine with tin(II) to form a tin(II) complex with the BPHA molecule, $(C_{13}H_{11}O_2N)_2SnCl_2$, m.p. 171°C. However, the precipitation titrations of tin(II) and tin(IV) with BPHA give tin to reagent mole ratios of 1 to 4 and 1 to 2, respectively. As both the products are the same [$Sn(BPHA)_2Cl_2$], it is suggested that BPHA acts as an oxidizing agent and not as a reducing agent. Thus 2 moles of BPHA are consumed in oxidizing tin(II) to tin(IV) before complexation.[3]

The infra-red spectral analysis shows[4] the tin complex to contain tin(IV). The compound is an inner complex of composition $(C_{13}H_{10}O_2N)_2SnCl_2$, and not an addition compound, since from the spectra, the tin compound appears to contain no hydroxyl hydrogen of BPHA and tin is combined with carbonyl oxygen.

The tin compound precipitates quantitatively from solutions that are 1–8% in conc. hydrochloric acid and can be weighed directly or ignited to oxide as the weighing form. However, though good results are obtained by igniting to oxide, it is more advantageous to weigh as the complex.

Tin in brass can be determined by precipitation from solutions containing 7% of conc. hydrochloric acid. Copper, lead or zinc remain in solution at that acidity. Copper, however, can be determined in the filtrate after raising its pH to 3·6–6·0 with ammonia.[1]

For complete precipitation, BPHA must be present in sufficient excess. But because of its light solubility, its addition must be closely regulated, particularly when the complex is to be weighed directly. The presence of more than 8% of concentrated hydrochloric acid gives low results. This is not due to the formation of $SnCl_6^{2-}$, because the precipitation is complete from solutions 15% in chloride.

Determination of Tin

(i) *By ignition to the oxide.* To a tin(IV) chloride soln. (containing 9·50 mg of Sn) add a few ml of conc. hydrochloric acid and dilute to 100 ml. The acidity of the soln. should be maintained between 1% and 5% in conc. acid. Mix slowly with stirring a 10% soln. of BPHA in ethanol (3 ml).

Cool in an ice bath for 1 hr, filter through a white ribbon filter paper (S & S 589), wash with cold water, ignite to the oxide and weigh.

Weight of tin oxide (SnO_2) × 0·78766 = weight of tin.

(ii) *By direct weighing of the complex.* Add conc. hydrochloric acid (10 ml) to the tin(IV) chloride soln. and dilute to 200 ml. Add dropwise with stirring a 1% soln. of BPHA in ethanol (5 ml for each 10 mg of tin present and then another 8 ml in excess). Cool in an ice bath for 4 hr, filter through a sintered glass crucible, wash with a few ml of ice-cold water, dry at 110° and weigh.

Weight of tin complex × 0·1927 = weight of tin.

DETERMINATION OF SCANDIUM AND SEPARATION FROM RARE EARTHS AND ZIRCONIUM
(Alimarin and Yun-hsiang[5])

Scandium is quantitatively precipitated from a solution at pH > 5·2 and thus can be separated from the rare earths. For the separation of scandium from zirconium, prior precipitation of the latter from 2 N hydrochloric acid is necessary. The precipitate, of composition $Sc(C_{13}H_{10}O_2N)_3$, is soluble in several organic solvents such as n-butanol, amyl acetate, benzene, chloroform and iso-amyl alcohol and on ignition at 600° is converted to the oxide, Sc_2O_3. The complex is stable up to 220°.

For the complete precipitation of scandium at the mentioned pH, a small excess of BPHA is required.

Determination of Scandium

Dilute the chloride soln. of scandium to 100 ml. Add to it a 20% ammonium acetate soln. (30 ml) and a few ml of 1:1 acetic acid to adjust the pH of the soln. to 5·4. Add a boiling aqueous soln. of BPHA in excess (Note 1) to precipitate scandium completely. Stir thoroughly for some time to coagulate the precipitate, filter through a quantitative filter paper, wash with water containing a little reagent, ignite to oxide at 600° and weigh.

Weight of scandium oxide (Sc_2O_3) × 0·6527 = weight of scandium.

Separation of Scandium from the Rare Earths

Dilute the soln. containing scandium and rare-earth elements to 100 ml (Note 2). Adjust the pH of the soln. to 5·4 by adding a 20% ammonium acetate soln. (30 ml) followed by a few ml of (1 + 1) acetic acid. Precipitate and determine scandium as described above.

From the combined filtrate and washings, precipitate the rare-earth elements as their hydroxides by adding ammonia, filter through a fine pore filter paper, wash with warm water, ignite at 800–900° and weigh.

Separation of Scandium from Zirconium

To the soln. of scandium and zirconium, add sufficient conc. hydrochloric acid so that after dilution to 100 ml the acidity of the soln. is 2 N. Dilute to 100 ml. Add a boiling aqueous soln. of BPHA, about double the theoretical amount. Filter off the zirconium precipitate and wash it with water. Evaporate the combined filtrate and washings to 100 ml. Neutralize the excess acid with ammonia, using methyl orange as indicator. Precipitate and determine scandium according to the procedure described above.

Notes

1. Preferably twice the theoretical amount.
2. Cerium(IV), if present, must be reduced to cerium(III) by boiling with hydroxylammonium chloride, otherwise scandium precipitated at pH 5·4 will be highly contaminated.

DETERMINATION OF COBALT AND NICKEL AND THEIR SEPARATION FROM COPPER
(Sinha and Shome[6])

Cobalt and nickel complexes precipitated from the hot solutions have the compositions $Co(C_{13}H_{10}O_2N)_2$ and $Ni(C_{13}H_{10}O_2N)_2$. The complexes can be weighed as such, or the cobalt complex may be converted to sulphate and that of nickel to oxide as the weighing form. Separation of cobalt and nickel from each other is not possible, but either of the metals is readily separated from copper. During the precipitation of cobalt, hydroxylamine must be added to prevent any formation of the cobalt(III) complex. The pH range for precipitation for both cobalt and nickel is 5·5–6·5. Above pH 6·5 the precipitates become hard lumps and also, in the case of nickel, the solution becomes coloured. Below pH 5·5 precipitation is incomplete.

Determination of Cobalt and Nickel

(i) *By ignition of the precipitates.* Dilute the soln. containing cobalt (26·2–31·5 mg) or nickel (24·4–30·5 mg) as its sulphate to 300 ml with distilled water and boil. If cobalt is to be determined, add a 2% soln. of hydroxylammonium chloride (10 ml). Add an ethanolic soln. of BPHA (0·25–0·30 g in 10–15 ml) slowly to the hot soln. and then a 10% sodium acetate soln. (5–10 ml) until the pH is 5·5–6·5. Stir occasionally on a boiling-water bath for 1 hr, filter, wash and ignite the precipitate to the oxide.

Nickel can be weighed as its oxide, but for cobalt, heat the oxide with a mixture of conc. nitric and sulphuric acids and weigh as the anhydrous cobalt(II) sulphate.

Weight of cobalt sulphate $\times$ 0·3802 = weight of cobalt.
Weight of nickel oxide (NiO) $\times$ 0·7858 = weight of nickel.

(ii) *By direct weighing of the complexes.* Precipitate the cobalt or nickel complexes as described above. Filter through a No. 3 sintered glass crucible, wash well with hot water, dry at 110–20° and weigh.

Weight of cobalt complex $\times$ 0·1220 = weight of cobalt.
Weight of nickel complex $\times$ 0·1216 = weight of nickel.

Separation of Cobalt or Nickel from Copper

Dilute the soln. containing copper (26·4–33·1 mg) and either cobalt (26·2–31·5 mg) or nickel (24·4–30·5 mg) as their sulphates to 400 ml with water. Add 1·5 N sulphuric acid (4–5 ml) and a 10% sodium acetate soln. (15 ml) to adjust the pH to 4. Heat the soln. to boiling and precipitate copper by the slow addition of BPHA [0·25–0·35 g in 5–10 ml of ethanol]. Heat the copper precipitate for 1 hr on a water bath, stirring occasionally. Filter through a No. 3 sintered glass crucible, wash well with water maintained at pH 4 by the addition of sufficient sodium acetate and sulphuric acid, and then with hot water. Dry at 110–20° to constant weight.

Weight of copper complex $\times$ 0·1302 = weight of copper.

Evaporate the combined filtrate and washings to about 300 ml, filter off any precipitated organic matter and then determine cobalt or nickel as described above. In the cobalt determination, add 20 ml of 2% hydroxylammonium chloride soln. before the addition of the precipitant.

SEPARATION AND DETERMINATION OF THORIUM AND CERIUM
(Sinha and Shome[7])

BPHA precipitates thorium and cerium(III) quantitatively in the pH ranges 4·5–5·5 and 6·5–7·5, respectively, Both elements may be determined after ignition of the precipitates to the respective oxides. Cerium(III) and thorium can be separated from each other by pH control.

Determination of Thorium

Dilute the soln. of thorium nitrate (containing 70–85 mg of ThO_2) to 250 ml with water. Adjust to pH 5 with a 10% ammonium acetate soln. Add BPHA (0·3–0·4 g) in ethanol (10–15 ml) with constant stirring. Continue stirring the white precipitate occasionally for a further 15 min, then filter, wash with water, ignite to thorium dioxide in a crucible and weigh.

Weight of thorium oxide (ThO_2) $\times$ 0·8788 = weight of thorium.

Determination of Cerium

Heat the soln. of cerium(IV) nitrate (containing 60–72 mg of CeO_2) to boiling, add a 5% hydroxylammonium chloride soln. (10 ml) to reduce the cerium(IV). Dilute to 250 ml with water, cool and add more (10 ml) of the hydroxylammonium chloride soln. to prevent oxidation, during precipitation, of the cerium(III). Add BPHA solution (0·3–0·4 g in 10–15 ml of ethanol) and then 1 N ammonia soln. to bring the pH to 7 to ensure complete precipitation. Stir the white precipitate occasionally for 30 min, filter, wash with water, ignite to cerium dioxide in a crucible and weigh.

Weight of cerium oxide (CeO_2) $\times$ 0·8141 = weight of cerium.

Separation of Thorium from Cerium

Heat the soln. to boiling and add hydroxylammonium chloride soln. as above. Cool the soln., dilute to 250 ml and add hydroxylammonium chloride soln. (10 ml). Raise the pH to 5 and precipitate **thorium** as described above. Filter, and wash the precipitate quickly with a freshly prepared 0·1% hydroxylammonium chloride soln. and finally with water. Dry, ignite the precipitate and weigh.

Combine the filtrate and washings and evaporate to 250 ml. Filter off the organic matter precipitated during evaporation, wash the filter and add more (10 ml) of the 5% hydroxylammonium chloride soln. to the filtrate and determine **cerium(III)** as described above.

DETERMINATION OF THORIUM AND SEPARATION FROM THE RARE EARTHS AND URANIUM
(Alimarin and Yun-hsiang[8])

The precipitation of thorium by BPHA begins at pH > 2; the precipitate, when dried at 105–20°, has the composition $Th(C_{13}H_{10}O_2N)_4$ and is stable up to 220°. By precipitation in an acetate buffer medium maintained approximately at pH 4·5, thorium can be separated from the rare earths. It can also be separated from uranium in the presence of ammonium carbonate at pH 7·0–8·5.

Determination of Thorium

To about 50 ml of thorium chloride soln. (containing 5·9–17·6 mg of ThO_2), add a boiling aqueous 0·4% soln. of BPHA (50 ml). Adjust the pH of the soln. to 4·5. Heat on a boiling water bath for 5–10 min so that the flaky precipitate formed at first becomes crystalline. Filter through a quantitative filter paper, wash with water containing a small quantity of the reagent, ignite to thorium dioxide at 800–900° and weigh.

Separation of Thorium from the Rare Earths

To 150 ml of soln. containing thorium and rare earths, add a 20% ammonium acetate soln. (50 ml) and a few ml of (1:1) acetic acid so that the pH of the soln. after precipitation is approximately 4·5. Proceed as above for the precipitation and determination of thorium.

Separation of Thorium from Uranium

Dilute the soln. to 150–200 ml and add a 20% ammonium carbonate soln. in sufficient quantity to keep the uranium in soln. Adjust the pH to 7·0–8·5. Precipitate thorium, filter, ignite and weigh as described above.

DETERMINATION OF THORIUM BY DIRECT WEIGHING AND SEPARATION FROM URANIUM, LANTHANUM, CERIUM, TITANIUM, ZIRCONIUM, GALLIUM, INDIUM, IRON, ALUMINIUM AND VANADIUM
(Das and Shome[9])

The dirty white thorium complex precipitated from hot solution (50–55°) at pH 4–5 has the composition $Th(C_{13}H_{10}O_2N)_4$ when dried at 105–10°. It is slightly soluble in water above 70° and readily soluble in 50% ethanol and also in other organic solvents such as chloroform, benzene, ether and ethyl acetate. The complex gives a gummy mass if the temperature of the solution in which it is formed is raised to over 80°. The amount of BPHA required for the complete precipitation of thorium at pH 4–5 is 2·0–2·5 times the stoichiometric amount.

Thorium has been separated from many other ions by the control of pH or by the use of masking agents. Thus thorium can be separated from cerium(IV) at pH 4·8 after the reduction of the latter with ascorbic acid, from lanthanum and uranium(VI) at pH 4·5 in presence of ammonium acetate, and from indium at pH 4·8 with thioglycollic acid as the masking agent. Prior precipitation, in presence of tartaric acid, is required for the removal of iron at pH 4·5, of aluminium at pH 5·1 and of vanadium(V) at pH 3·8. Titanium[22] or zirconium[11], if present, is precipitated from 0·5 N sulphuric acid and gallium[15] is precipitated at pH 2, before thorium is determined in the filtrate.

Phosphate, fluoride, citrate and EDTA must be absent, as they interfere with the precipitation of thorium.

Determination of Thorium

Dilute the nitrate soln. containing thorium to 200 ml. Heat to 50–55° and add an ethanolic soln. of BPHA (8–10 ml, 2·0–2·5 times the stoichiometric amount) followed by a 10% ammonium acetate soln. until the pH is 4·5. Digest the precipitate at 50–55° on a water bath until the soln. becomes clear and the precipitate is granular (Note). Filter through a sintered glass crucible, wash with water at 40–45°, dry at 110° for 1 hr and weigh.

Weight of thorium complex $\times$ 0·2147 = weight of thorium.

Note. The time required for the complete granulation of the precipitate and the soln. to become clear [is more than 2 hr. If, however, 1–2 g ammonium chloride are added before the adjustment of pH, the precipitate coagulates quickly into an easily filterable form.

Separation of Thorium from Uranium, Lanthanum and Cerium

Add ascorbic acid in sufficient amount to reduce cerium(IV) to cerium(III), heat the soln. to boiling, then dilute to 200 ml. Heat to 50–55°, add an excess of BPHA soln., adjust the pH to 4·8 with 10% ammonium acetate soln. and then proceed as above for the determination of thorium.

The same procedure as suggested for the determination of thorium when present alone, may be followed for its separation from uranium(VI) and lanthanum.

Separation of Thorium from Titanium or Zirconium

Dilute the soln. to 250 ml. Add dilute sulphuric acid to make the soln. 0·5 N in acid and then add the required amount of BPHA (about 2·5–3·0 times the stoichiometric amount) in ethanol (10 ml), to precipitate completely zirconium or titanium. Stir occasionally and digest the precipitate on a steam bath. Filter and wash the precipitate with water containing a little BPHA.

Concentrate the combined filtrate and washings to about 200 ml, filter off any organic matter, adjust the pH to 4–5 by adding ammonium acetate, and precipitate **thorium** as described above.

Separation of Thorium from Gallium

Neutralize the soln. with dilute ammonia soln., and add dilute hydrochloric acid so that on dilution to 200 ml the pH of the soln. is 2. Dilute to 200 ml, heat to 40–50° and add BPHA soln. (10 ml of ethanol containing 2·5–3·0 times the stoichiometric amount of BPHA). Heat at 60–70° on the water-bath with occasional stirring until the supernatant soln. is clear. Filter and wash the **gallium** precipitate with hot water.

Concentrate the filtrate and washings to 200 ml and filter off any separated organic matter. Determine thorium in this soln. according to the procedure described above.

Separation of Thorium from Vanadate

Add sufficient tartaric acid to complex the metal ions present. Dilute to 250 ml, add 10% sodium acetate soln. to adjust the pH to 3·8, heat to 50° and add BPHA (about twice the stoichiometric amount in 10 ml of alcohol) slowly with stirring. Keep it at 60° on the water bath for some time and stir thoroughly. Filter the vanadium precipitate and wash with water at pH 4 containing a small amount of the reagent.

Concentrate the filtrate and washings to 200 ml, filter off any organic matter and heat to 50–55°. Add a requisite amount of the BPHA in ethanol, adjust the pH to 6·2 with a 10% sodium acetate soln. and determine thorium as described above.

Separation of Thorium from Iron or Aluminium

Neutralize the soln. with ammonia. Add 5 ml of 1 N sulphuric acid and enough tartaric acid to complex the iron or aluminium. Dilute with water to 400 ml and heat to 60°. Add an ethanolic soln. of BPHA (about 2 times the stoichiometric amount in 10 ml) and a 10% sodium acetate soln. to raise the pH to 4·5 for iron or to 5·1 for aluminium. Heat at 65° on a water bath for 2 hr. Filter and wash the precipitate of iron or aluminium complex with hot water containing 0·1% reagent and adjust to the pH of precipitation of the elements.

Concentrate the filtrate and washings to 200 ml, filter off organic matter, precipitate and determine thorium by the addition of the requisite amount of BPHA, adjusting the pH to 6·2 with the sodium acetate soln.

Separation of Thorium from Indium

Add sufficient thioglycollic acid to keep indium in soln. Dilute to 200 ml, heat to 50° and add BPHA soln. (2·5 times the stoichiometric amount in 10 ml of ethanol). Raise the pH to 4·8 by addition of 10% sodium acetate soln. Digest the precipitate and determine thorium as described above.

DETERMINATION OF ZIRCONIUM AND SEPARATION FROM OTHER IONS
(Alimarin and Yun-hsiang[10])

This method proposes the use of a large excess of BPHA for the quantitative precipitation of zirconium from solutions 2·4 N in hydrochloric acid or 3·6 N in sulphuric acid. As tartaric acid (5%) and hydrogen peroxide (0·15%) have no effect on the precipitation of zirconium, they can be used to prevent the precipitation of titanium, vanadium, niobium, tantalum, iron, aluminium, chromium and the rare earths. The zirconium precipitate, after filtration and washing, is ignited to the oxide as the final weighing form. However, the precipitate obtained from > 3 N hydrochloric acid solution has the composition $Zr(BPHA)_4$. From the thermolysis curve, the complex appears to be stable up to 240° and is converted completely to oxide at 500°.

Determination of Zirconium

To about 150 ml of zirconium soln. (containing 1·6–30·3 mg of ZrO_2) add enough conc. hydrochloric of sulphuric acid so that the soln. when finally diluted to 250 ml is either 2·4 N in hydrochloric acid or 3·6 N in sulphuric acid. Add BPHA (0·4–0·5 g) dissolved in boiling water (100 ml) to precipitate zirconium completely. Heat the precipitate on a boiling water bath for 15–20 min with occasional stirring. Filter immediately through a quantitative filter paper, wash with 1% hydrochloric acid in 0·05% BPHA, dry and ignite in a porcelain crucible at 1000–1100° to constant weight.

Weight of zirconium oxide (ZrO_2) $\times$ 0·7403 = weight of zirconium.

Determination of Zirconium in the Presence of Titanium, Niobium, Tantalum, Vanadium, Iron, Aluminium and Chromium

To about 150 ml of the soln. containing sufficient tartaric acid to give a concentration of 5% to the total volume of 250 ml, add with stirring conc. sulphuric acid (25 ml) (Note) followed by a 30% hydrogen peroxide soln. (1 ml) and BPHA (0·4–0·5 g) dissolved in boiling water (100 ml). Digest the precipitate on a simmering water bath for not more than 15 min, filter, wash and ignite the precipitate as described above.

Note. Precipitation of vanadium occurs from 2·4 N hydrochloric acid thus preventing its separation from zirconium under the condition.

DETERMINATION OF ZIRCONIUM
(Ryan[11])

BPHA precipitates zirconium from sulphuric or hydrochloric acid solutions. When the reagent is used only in slight excess over the stoichiometric amount required for the zirconium present, the acidity of the solution for the complete precipitation of zirconium must not be greater than 0·5–0·6 N (5% of concentrated hydrochloric acid or 1·5% of concentrated sulphuric acid). But from hydrochloric acid solutions, the precipitate obtained is not of any definite composition. It must, therefore, be ignited to the oxide as the weighing form. The complex formed in sulphuric acid solution, however, has the composition $Zr(C_{13}H_{10}O_2N)_4$ when dried at 110°, and hence can be weighed as such or ignited to the oxide for the gravimetric determination of the element. For the complete precipitation of very small amounts of zirconium, a prolonged or even an overnight digestion period is necessary. The method is very sensitive. Even 1 ppm of zirconium can be detected and as little as 0·2 mg of zirconium can be determined. During the precipitation of zirconium from hydrochloric acid, copper, aluminium, bismuth, cadmium, chromium(III), lead, magnesium, manganese, nickel, zinc, uranyl, thorium, yttrium, cerium(III), samarium(III), lanthanum, praseodymium, gadolinium, neodymium and dysprosium do not interfere. The interference of titanium(IV) and cerium(IV) is largely prevented by the addition of hydrogen peroxide.

Determination of Zirconium

(i) *By ignition to the oxide.* To the zirconium soln. diluted to 150–200 ml and containing up to 5% of conc. hydrochloric acid or 1·5% of conc. sulphuric acid, add BPHA (4% in 95% ethanol) in excess of the stoichiometric quantity required for the zirconium present. Digest for 30 min on a boiling water bath. After cooling to room temperature, filter through a Whatman No. 42 filter paper, wash with a saturated aqueous reagent soln. and ignite to ZrO_2.

(ii) *By direct weighing of the precipitate.* Dilute the zirconium sulphate soln. to 100–150 ml and add sufficient sulphuric acid to make the soln. 0·5 N in acid. Add dropwise, with stirring, a 4% BPHA soln. in 95% ethanol in slight excess (not more than 40 mg per 100 ml of the soln.) over the stoichiometric quantity required for the zirconium present. Digest on the steam bath for 1 hr. Cool, filter through a porous porcelain crucible, wash with cold water and finally 2 to 3 times with 5-ml portions of hot water (80°). Dry at 110° and weigh.

Weight of zirconium complex × 0·0970 = weight of zirconium.

DETERMINATION OF MOLYBDENUM AND SEPARATION FROM COPPER, COBALT, NICKEL, IRON AND VANADIUM
(Sinha and Shome[12])

Molybdenum is quantitatively precipitated at below 70° from 0·01 to 2·5 N hydrochloric acid solutions when the supernatant liquid contains 0·03% of BPHA in excess. The precipitate is soluble in concentrated acid, in ethanol and slightly in water at or above 80%. Because the precipitate, of composition $MoO_2(C_{13}H_{10}O_2N)_2$, decomposes only at 165–6°, it can be weighed directly after drying or it can be converted to the oxide, MoO_3, as the weighing form. Molybdenum has been determined in the presence of appreciable amounts of iron(III), copper(II), nickel(II) and cobalt(II) and much smaller amounts of chromium(VI) and vanadium(V). While copper, cobalt, nickel and chromium(VI) remain in solution in the acidity required for molybdenum precipitation, iron(III) and vanadium(V) must be masked, for example by EDTA.

Determination of Molybdenum(VI)

(i) *By ignition to the oxide.* Dilute the soln. of molybdate (containing 77 mg of Mo) to 300 ml with water. Add sufficient 10 N hydrochloric acid to keep the acidity between 0·01–2·5 N and then add slowly with stirring 10–15 ml of an ethanolic soln. containing 0·44 g of BPHA. The amount of BPHA required is the stoichiometric quantity plus at least 30 mg per 100 ml of soln. The final soln. must not contain more than 5% of ethanol. Stir the yellow precipitate that appears for about 15 min and allow the precipitate to settle for another 15 min. Filter, wash the precipitate with 0·02 N hydrochloric acid, neutralize the free acid, if any, in the precipitate with one or two drops of 6 N ammonia solution, ignite at 500–25° and weigh as molybdenum trioxide.

Weight of molybdenum oxide (MoO_3) × 0·6665 = weight of molybdenum.

(ii) *By direct weighing of the precipitate.* Dilute the molybdate soln. (containing 30–77 mg of Mo) to 300 ml with water. Add 10 N hydrochloric acid (1 ml), heat to 60° and precipitate molybdenum by adding BPHA

(0·25–0·44 g) in ethanol (10–15 ml). Heat at 60° on a water bath for 30 min, filter through a No. 3 sintered glass crucible, wash with 0·02 N hydrochloric acid at 60° and finally with water (10 ml), dry for 1 hr at 110–15° and weigh.

Weight of molybdenum complex × 0·1737 = weight of molybdenum.

Separation of Molybdenum(VI) from Copper, Nickel, Cobalt or Chromium(VI)

Add 6 ml of 10 N hydrochloric acid to the test soln., dilute to 300 ml, heat to 60° and precipitate molybdenum as described above. Filter, wash the precipitate with warm (60° 0·2 N) hydrochloric acid followed by warm 0·02 N hydrochloric acid and finally with a few ml of water. Dry the precipitate at 110° and weigh.

When chromium(VI) is present, wash the precipitate with 0·02 N hydrochloric acid, then with water and ignite it to the oxide.

Separation of Molybdenum from Iron(III) or Vanadate

Acidify the 300 ml of molybdate soln. containing iron (384–772 mg) or vanadate (4–8 mg) (Note) with 10 N hydrochloric acid, adding 0·4 ml to the soln. containing iron and 0·8 ml to the soln. containing vanadate. Add disodium EDTA (3–6 g for iron or 2–3 g for vanadate), heat to 60° and precipitate molybdenum, as described above. Filter, wash with warm (60°) 0·02 N hydrochloric acid and then with water (10–15 ml). Dry and weigh the precipitate.

Note. When vanadate is present, the soln. after the addition of EDTA must be heated for a few minutes to reduce vanadium(V) to vanadium(IV) before the procedure described above is followed.

DETERMINATION OF BERYLLIUM AND ITS SEPARATION FROM IRON, ALUMINIUM AND TITANIUM
(Das and Shome[13])

The beryllium BPHA complex, having the composition $Be(C_{13}H_{10}O_2N)_2$, is quantitatively precipitated between pH 5 and 8. The metal may be determined either by igniting the precipitate to oxide or by weighing the complex. The white granular precipitate is not soluble in water if the temperature is not above 70°. It is soluble in chloroform but is decomposed by moderately concentrated mineral acids. It melts with decomposition at 220°. Beryllium may be separated from iron, aluminium and titanium by the prior precipitation of the first two at pH 4, and titanium at about pH 1; beryllium is then determined in the filtrate.

For the complete precipitation of beryllium, the supernatant solution must contain 0·1% of BPHA in excess.

Determination of Beryllium

(i) *By ignition of the precipitate.* Dilute the soln. containing 3 to 6 mg of Be to 200 ml, heat to 50–60° and add the ethanolic BPHA soln.(10 ml) slowly with constant stirring. About 2·0–2·5 times the stoichiometric amount of BPHA is required (Note 1). Add 3 N ammonia soln. slowly to raise the pH of the soln. to 5·5–6·5. Allow to stand for a few min, filter, wash with warm water, ignite and weigh.

Weight of beryllium oxide (BeO) × 0·3605 = weight of beryllium.

(ii) *By direct weighing of the precipitate.* Precipitate beryllium as above, heat to 50–60° for 15 min and stir well. Filter through a No. 3 sintered glass crucible, wash with water (50–60°), dry at 110° and weigh.

Weight of beryllium complex × 0·0208 = weight of beryllium.

Separation of Beryllium from Aluminium and Iron(III)

Neutralize the soln. of beryllium containing aluminium (< 34 mg) or/and iron(III) (< 38 mg) with 2 N ammonia soln. and acidify with a few drops of dilute sulphuric acid. Dilute to 300 ml. Add a 10% sodium acetate soln. to adjust the pH to 4. Heat to 65° and add BPHA (0·4–0·9 g in 10–15 ml of ethanol) to precipitate aluminium and/or iron (Note 2). Allow the precipitate to stand at 65–70° on a water bath (Note 3) for 90 min with occasional stirring. Filter and wash with hot water adjusted to pH 4 and containing 0·1 g of BPHA per 100 ml.

Evaporate the filtrate and washings to 200 ml and filter off any organic matter. Add BPHA (0·10–0·15 g, if required, dissolved in 5 ml of ethanol) for the complete precipitation of beryllium followed by 3 N ammonia soln. to adjust the pH to 5·5–6·5. Digest for 15 min at 50–60° with occasional stirring, filter, wash, dry and weigh as $Be(C_{13}H_{10}O_2N)_2$ according to the procedure described above.

Separation from Titanium

Dilute the soln. containing beryllium and titanium (2–17 mg) to 300 ml, neutralize with 6 N ammonia soln. and acidify with conc. hydrochloric acid (3–4 ml). Precipitate titanium by adding BPHA (0·4–0·7 g dissolved in 10 ml of ethanol) with stirring (Note 4). Allow to stand for 45 min and stir occasionally. Filter, wash with BPHA (1 g) in ethanol (10 ml) and conc.

hydrochloric acid (3 ml) dissolved in water (1 litre). Collect the filtrate and washings, concentrate and determine beryllium by the addition of a fresh quantity of BPHA as above.

Notes

1. At least 0·1 g of BPHA per mg of beryllium is required.
2. For the complete precipitation of iron and aluminium at pH 4, about twice the stoichiometric amount of BPHA is required.
3. The iron precipitate becomes gummy if the temperature is raised above 70°.
4. For the complete precipitation of very small quantities of titanium from acidic soln. a large excess of BPHA is required.

DETERMINATION OF URANIUM(VI) AND SEPARATION FROM LEAD, BISMUTH, THORIUM, CERIUM, IRON, ALUMINIUM, TITANIUM, ZIRCONIUM AND MOLYBDENUM
(Das and Shome[14])

BPHA precipitates uranium(VI) quantitatively between pH 5·2 and 5·6. The orange-red precipitate changes on digestion to a less soluble brick-red granular form which can be weighed by ignition to the oxide, U_3O_8 or as the complex $UO_2(C_{13}H_{10}O_2N)_2$. The freshly precipitated flocculent mass, which shows some solubility in water above 60°, is less soluble on granulation. It is soluble in chloroform, ether and acetone, but less so in carbon tetrachloride. The solution during precipitation must not contain more than 20% of ethanol, for at higher concentrations, the uranium complex has some solubility. The complex decomposes on treatment with strong mineral acids or caustic soda and by dry heat at 204–6°.

Iron(III), titanium(IV), zirconium, molybdenum(VI) and small quantities of aluminium may be removed by their prior precipitation with BPHA by proper adjustment of the acidity. For instance, iron is removed by precipitation at pH 3·0–3·5, titanium, zirconium and molybdenum(VI) are precipitated from hydrochloric acid solutions of normality 0·15, 2·4 and 0·01–2·5, respectively. For the complete precipitation of aluminium at pH 4 a large excess of BPHA is necessary which may, at the same time, cause the precipitation of some uranium. Therefore prior precipitation and separation of aluminium is possible only when it is present in relatively small amounts.

Interference of cerium(III), thorium, lead and bismuth is masked by addition of magnesium-EDTA soln.; EDTA alone incompletely precipitates uranium.

Organic acids, carbonate and fluoride, which hamper the precipitation of uranium(VI), must be absent and so also must beryllium, as it precipitates along with the uranium. Chromium(III), vanadium(V) and tungsten(VI) are not masked by magnesium-EDTA. Chromium(VI) oxidizes the reagent.

The least quantity of BPHA required for the complete precipitation of uranium is about 3 times the stoichiometric amount.

Determination of Uranium

(i) *By ignition of the precipitate*. Dilute the soln. (containing 31–44 mg of uranium) to 200 ml, heat to 40–50° and add slowly with stirring a soln. of BPHA (0·3 g) in ethanol (10 ml). The soln. must contain at least 3 times the stoichiometric quantity of BPHA. Add a 2 N ammonia soln. to raise the pH to 5·4. Keep at 40–50° on a water bath for 90 min and stir occasionally. Ensure that the pH is maintained, if need be, by the addition of one or two drops of ammonia. Filter through a quantitative filter paper, wash with water at 40–50°, dry and ignite. Dissolve the ignited oxide in the crucible with a few drops of conc. nitric acid, evaporate and reignite.

Weight of uranium oxide (U_3O_8) × 0·8480 = weight of uranium.

(ii) *Determination by direct weighing of the precipitate*. Precipitate the uranium as described above. Filter through a No. 3 sintered glass crucible, wash with water at 40–50°, dry at 110° and weigh.

Weight of uranium complex × 0·3428 = weight of uranium.

Determination of Uranium in the Presence of Lead, Bismuth, Thorium and Cerium(III)

To the soln. of uranium containing cerium(III), thorium, lead or bismuth (25–50 mg) add a magnesium–EDTA soln. (20 ml of 0·25 M EDTA soln. + 5 ml of 1 M magnesium chloride soln.), dilute to 300 ml and heat to 40–50° on water bath. Add with stirring BPHA (0·3 g) dissolved in ethanol (10 ml), adjust to pH 5·4 by the dropwise addition of dilute ammonia, digest the precipitate and determine uranium as described above.

Determination of Uranium in the Presence of Aluminium, Iron, Titanium, Zirconium or Molybdenum(VI)

Dilute the test soln. to 300 ml and add conc. hydrochloric acid or 2 N ammonia soln. to maintain the acidity at 0·15 N for titanium, 2·4 N for zirconium, 1 N for molybdenum, pH 3·5 and 4·0 for iron and aluminium, respectively.

Precipitate titanium (in the cold) with a sufficient amount of BPHA, wait for 1 hr, filter and wash. When one of the other ions is present, heat the soln. to 50–60°, add BPHA in sufficient excess for the complete precipitation of that ion and digest on a hot-water bath with occasional stirring. The digestion period for iron and aluminium is 90 min, while for zirconium and molybdenum the required period is only 15–30 min. Filter and wash the precipitate with water having the same acidity as that of the soln. from which the precipitation has been carried out and containing BPHA (0·1 g per 100 ml).

Combine the filtrate and washings and evaporate to 200 ml, filter off any organic matter, precipitate uranium with the addition of more reagent, adjust the pH of the solution and proceed as described above.

THE DETERMINATION OF INDIUM AND GALLIUM, THEIR SEPARATION FROM EACH OTHER AND FROM OTHER IONS
(Das and Shome[15])

Indium and gallium precipitate as white complexes with BPHA. The precipitation of indium begins at pH 3·9 and is complete between pH 4·3 and 8·0. Gallium starts to precipitate at pH 1; the precipitate has the formula $Ga(C_{13}H_{10}O_2N)_3$ when formed between pH 1·6 and 3·5. The precipitates may be weighed directly as $In(C_{13}H_{10}O_2N)_3$ and $Ga(C_{13}H_{10}O_2N)_3$, or after ignition to the respective oxides. The indium–BPHA precipitate is very slightly soluble in water above 70° and in ethanol and carbon tetrachloride, but very soluble in chloroform. It decomposes at 208–9°. The gallium–BPHA complex is very slightly soluble in carbon tetrachloride, benzene, ether and ethanol but is very soluble in chloroform. It decomposes at 226°.

Iron(III) and copper are separated from indium by their prior precipitation with the reagent from weakly acidic solutions in the presence of tartrate. Tin, if present, is removed at a much lower pH before the indium determination. While smaller quantities of zinc do not coprecipitate with indium, larger amounts of zinc, copper or nickel must be masked with cyanide. However, gallium can be determined in presence of copper, aluminium, thorium, cerium(III) and uranium(VI) simply by pH control, but titanium needs prior precipitation from 0·2 N acid.

As the pH for complete precipitation of gallium is much lower than that for indium, the former can easily be separated from the latter. The precipitation of gallium, like that of indium, should be made below 70° and its period of digestion should not be more than 45 min, to avoid any possibility of decomposition of the gallium complex. The amount of BPHA required for complete precipitation from a 200-ml solution of indium at pH 5·3–5·5 and of gallium at pH 2 is about 2·5 times the theoretical amount.

A large excess of iron(III) and fluoride prevents the complete precipitation of both gallium and indium; while phosphate does not interfere with the precipitation of gallium, indium is not quantitatively precipitated in its presence. Aluminium also affects the precipitation of indium. Tartrate, however, even in considerable amount, has no effect on the precipitation of both the elements.

Determination of Indium

(i) *By ignition of the precipitate.* Dilute indium sulphate soln. (containing 7–28 mg of indium) to 200 ml, adjust its pH to 4·8–5·3 (Note 1) with a 10% sodium acetate soln. and glacial acetic acid. Heat the soln. to 40–50° and add BPHA (0·2–0·5 g in 8–10 ml of ethanol) dropwise to precipitate indium. (The amount of BPHA required for complete precipitation is about 14 times that of the indium present.) Stir the precipitate for 10–15 min, digest at 60–70° on a water bath for 45–60 min, filter, wash with hot water, ignite and weigh.

Weight of indium oxide (In_2O_3) × 0·8271 = weight of indium.

(ii) *By direct weighing of the precipitate.* Precipitate the metal complex as described above. Heat at 60–70° on a water bath for 45–60 min with occasional stirring, filter through a No. 3 sintered glass crucible, wash with hot water, dry at 110–20° and weigh.

Weight of indium complex × 0·1528 = weight of indium.

Determination of Gallium

(i) *By ignition of the precipitate.* Adjust the pH of the gallium chloride soln. (containing 5–23 mg of gallium) diluted to 200 ml to 2·5–3·0 with hydrochloric acid. Heat the soln. to 40–50° and precipitate gallium by adding slowly a soln. of BPHA (0·2–0·5 g in 5–10 ml of ethanol) with stirring. (The amount of BPHA required for the complete precipitation is about 23 times that of the gallium present.) Stir for some time and digest at 60–70° on a water bath until the supernatant liquid becomes clear. Filter, wash the precipitate with hot water, ignite and weigh.

Weight of gallium oxide (Ga_2O_3) × 0·7439 = weight of gallium.

(ii) *By direct weighing of the precipitate.* Precipitate the gallium complex as described above, heat at 60–70° on a water bath for 30–45 min. Filter through a No. 3 sintered glass crucible, wash the precipitate, dry at 110–20° and weigh.

Weight of gallium complex × 0·0987 = weight of gallium.

Separation and Determination of Gallium and Indium

Dilute the soln. to 300 ml, adjust to pH 2·5, and precipitate and determine gallium as described above. For indium (14 mg), concentrate the filtrate along with the washings by evaporation to 250 ml, filter free of organic matter and adjust the pH to 4·8–5·3 with a 10% sodium acetate soln. and glacial acetic acid. Heat to 40–50°, precipitate indium by adding BPHA (0·2 g) in ethanol (8–10 ml) and determine indium as the complex.

Separation of Indium from Iron or Copper

Dilute the mixture containing indium (14 mg) and iron (16 mg) or copper (7 mg) to 300 ml with water, add sodium potassium tartrate (3 g) to prevent the precipitation of indium during the addition of BPHA. Adjust with dilute hydrochloric acid the pH of the soln. to 3·5 (for iron) or to 3·8 (for copper), heat to 65° and add dropwise with stirring a soln. of BPHA (0·5–1·0 g in 10–15 ml of ethanol) to precipitate iron or copper. Digest the precipitate on the water bath for 60–90 min, with occasional stirring. Filter, wash the precipitate with a hot 0·1% aqueous soln. of the reagent containing some sodium potassium tartrate and maintained at pH 3·5–3·8 by addition of hydrochloric acid. Reject the precipitate.

Concentrate the combined filtrate and washings to 250 ml, filter and add 3 N ammonia soln. to raise the pH to 5·5–6·0. Precipitate indium by a further addition of BPHA (0·2 g in 5 ml of ethanol) as already described and determine indium as $In(C_{13}H_{10}O_2N)_3$.

Separation of Tin from Indium

Add dilute hydrochloric acid to adjust the soln. containing indium (12 mg) and tin(II) (14 mg) ions and diluted to 200 ml to pH 0·5–1·0. Add BPHA (0·5–0·8 g in 8–10 ml of ethanol) slowly with stirring to precipitate the tin. Cool in cold water for a few hr with occasional stirring, filter, wash with cold aqueous 0·2% reagent soln. adjusted to pH 0·5–1·0 with hydrochloric acid, and reject the precipitate.

Evaporate the combined filtrates to 300 ml, filter free from organic matter, and raise the pH to 5·3–5·5 by the addition of a 10% sodium acetate soln. (Note 1). Add BPHA (0·2 g in 5 ml of ethanol) and determine indium by direct weighing of the complex, as described above.

Separation of Indium from Nickel, Copper or Zinc

Dilute the soln. containing indium (8–14 mg) and nickel, copper or zinc to 100 ml. Add sodium potassium tartrate (2 g) to keep indium in solution, followed by a 20% potassium cyanide soln. (Note 2). Adjust the pH to 5·8–6·6 and precipitate indium by gradually adding BPHA soln. (0·20–0·25 g of BPHA in 8–10 ml of ethanol) and weigh the complex as described above.

Separation of Gallium from Aluminium, Copper, Thorium, Uranium or Cerium

Dilute the soln. containing gallium (5–8 mg) and aluminium, copper, uranyl, thorium or cerium(III) ions to 300 ml and add dilute hydrochloric acid to adjust the pH to 2. Heat to 50° and add dropwise with stirring soln. of BPHA (0·3–0·5 g in 8–10 ml of ethanol) till the precipitation of gallium is complete. Digest on the water bath for 30–45 min, filter, wash, dry and weigh as $Ga(C_{13}H_{10}O_2N)_3$.

Separation of Gallium from Titanium

Add conc. hydrochloric acid to the 300 ml of soln. containing titanium (3·5–7·5 mg) and gallium to make the soln. 0·2 N in acid. Precipitate titanium by the gradual addition with stirring of BPHA (0·5–1·0 g in 8–10 ml of ethanol). Allow the precipitate to stand for 40–50 min with occasional stirring, filter, and wash with a cold aqueous 0·2% reagent soln. adjusted to pH 0·5. Evaporate the filtrate to 300 ml, filter off any separated organic matter, adjust the acidity of the filtrate to pH 2·5–3·0 and precipitate and determine gallium by direct weighing of the complex, as described above.

Notes

1. In the absence of sodium potassium tartrate, it is not advisable to take the pH of precipitation of indium above 5, because in the absence of tartrate, indium hydrolyses above pH 5.
2. The potassium cyanide must be in reasonable excess, particularly when zinc is present.

THE DETERMINATION OF GALLIUM IN THE PRESENCE OF ALUMINIUM, BERYLLIUM, IRON AND ZINC
(Alimarin and Hamid[16])

As gallium precipitates completely from perchloric acid at pH 1·5 and above and at about pH 2 from hydrochloric acid, a method for the determination and separation of gallium from aluminium, beryllium, zinc and iron, which do not precipitate at these acidities, has been suggested.

Procedure

Follow the method described above, adjusting the pH of the hydrochloric acid soln. to 2·3 and adding enough of 2% ascorbic acid soln. to keep iron in the iron(II) state. If the gallium precipitate after ignition to oxide is found to be contaminated with iron, dissolve the precipitate in conc. hydrochloric acid and reprecipitate.

DETERMINATION OF TUNGSTEN AND SEPARATION FROM MOLYBDENUM, URANIUM, TITANIUM, VANADIUM AND IRON
(Kaimal and Shome[17])

Tungsten precipitates completely from solutions 0·05 N to 1 N with respect to hydrochloric acid on the addition of about 3 times the stoichiometric amount of BPHA. In the presence of tartrate, however, the precipitation is complete only when the acidity is raised to 0·5 N. The precipitate first formed in the absence of tartrate is yellow, but it changes to a white crystalline form on digestion on a water bath. The white precipitate has the composition $WO_2(C_{13}H_{10}O_2N)_2$; it decomposes at 194–5° and is soluble in chloroform and in 95% ethanol. In the presence of tartrate, the precipitate is white even when formed in the cold.

Because the precipitation of tungsten, in presence of tartrate, does not begin at an acidity lower than 0·12 N, the separation of vanadium(V), titanium(IV), molybdenum(VI) and iron(III) from tungsten can be undertaken by the prior precipitation of vanadium(V) and titanium(IV) from a cold soln. at pH 1·0–1·5, and of molybdenum(VI) and iron(III) from hot solution at pH 1·0–1·5 and 4·0, respectively. Tungsten is determined in the filtrate. Tungsten is separated from uranium(VI) at the pH of its complete precipitation in the presence or absence of tartrate, as uranium does not precipitate even below pH 3.

Determination of Tungsten (and Separation from Uranium)

Dilute the tungstate soln. (containing 35 mg of tungsten) to 200 ml, heat to 50° and add a 3% BPHA soln. in ethanol (10 ml) slowly with stirring. Adjust the acidity to 0·1 N with 3 N hydrochloric acid, when a yellow precipitate separates. Heat on a water bath at exactly 60° for 45 min, while stirring the soln. frequently. Allow the white crystalline precipitate thus obtained to settle and cool to room temperature. Filter the precipitate through a sintered glass crucible, wash with water containing 3 N hydrochloric acid (2 ml) per 100 ml at 45–50°, dry at 115° for 1 hr and weigh as $WO_2(C_{13}H_{10}O_2N)_2$. Alternatively ignite at 900° and weigh as the oxide $(WO)_3$.

Weight of tungsten complex × 0·2872 = weight of tungsten.
Weight of tungsten oxide × 0·7930 = weight of tungsten.

Determination of Tungsten(VI) in the Presence of Tartrate

Dilute the tungstate soln. (containing 35 mg of tungsten) containing 10% sodium potassium tartrate soln. (10 ml) to 200 ml. Add BPHA (0·3 g in 20 ml of ethanol) slowly with stirring and adjust the acidity to 0·5–1 N with 6 N hydrochloric acid. Thoroughly stir the precipitate and heat on the water bath at 60° for 1 hr. Cool to room temperature, filter, wash and weigh as described above.

Separation of Tungsten from Vanadium and Titanium

Dilute the test soln. to 200 ml, add 10% sodium potassium tartrate soln. (10 ml), adjust the pH to 1·0–1·5 and precipitate vanadium and titanium by adding BPHA (twice the stoichiometric amount in 8 ml of ethanol). Stir the precipitate, filter through a quantitative filter paper and wash several times with a dilute BPHA soln. (0·1 g of BPHA in 3–4 ml of ethanol diluted with water to 100 ml) having an acidity the same as that of the precipitation soln.

Evaporate the combined filtrate and washings to 200 ml. Filter the soln. free from any suspended organic matter and wash the filter with water. Precipitate and determine tungsten according to the procedure described above for the determination of tungsten in presence of tartrate.

Separation of Tungsten from Molybdenum and Iron

Add to the soln. (200 ml) containing tungstate, molybdate and iron(III), 10% sodium potassium tartrate soln. (10 ml). Adjust the pH to 1·0–1·5 and heat on a water bath to 80°. Precipitate molybdenum by adding BPHA (2 times the stoichiometric amount in 8 ml of ethanol). Allow the precipitate to settle at 80° for some time. Filter the precipitate through a quantitative filter paper and wash several times with the wash soln. (see immediately above) maintained at the pH of precipitation.

Combine the filtrate and washings, adjust to pH 4 with dilute ammonia solution, heat to 65° and precipitate iron by adding twice the stoichiometric amount of BPHA in ethanol (10 ml). Allow the precipitate to settle at 65°

for some time, filter through a quantitative filter paper and wash the precipitate with the same wash soln. as mentioned before, but maintained at pH 4.

Evaporate the combined filtrate and washings to 200 ml. Filter off any separated organic matter and wash with water. Precipitate and determine tungsten as described above.

DETERMINATION OF LANTHANUM AND SEPARATION FROM URANIUM, CERIUM, IRON, ALUMINIUM, THORIUM, GALLIUM, INDIUM, SCANDIUM, ZINC, NICKEL AND COPPER
(Das and Shome[18])

BPHA has been proposed as a reagent for the gravimetric determination of lanthanum. It precipitates the element quantitatively between pH 6·4 and 7·2. The precipitate, of composition $La(C_{13}H_{10}O_2N)_3$, may be weighed as such or may be ignited to the oxide La_2O_3 as the weighing form. The precipitate, which decomposes at 135–6°, is soluble in mineral acids and in organic solvents such as chloroform, ether, benzene and ethanol; it is sparingly soluble in acetone, acetic acid and ethyl acetate and slightly so in water above 60°.

For the complete precipitation of lanthanum from 200 ml of solution, 2·00–2·25 times the stoichiometric amount of BPHA is required. The optimum temperature is 50°. The period of digestion of the precipitate must not be prolonged as otherwise the complex may decompose. Because iron(III), aluminium, gallium, indium, thorium, scandium, cerium(IV) and uranium(VI) precipitate, respectively, at pH 3·5, 4·0–4·5, 2·0, 4·5, 4·5–5·0, 5·2–5·4, 4·8–5·2, 5·2–5·6, these elements can be separated from lanthanum by their prior precipitation. Nickel, copper and zinc and small amounts of iron(II) interfere, but may be masked by cyanide. Citrate or EDTA prevents the precipitation of lanthanum while tartrate or carbonate gives incomplete precipitation of the element. Phosphate and fluoride also interfere seriously.

Determination of Lanthanum

Dilute the soln. (containing 5–26 mg of lanthanum) to 200 ml. Heat to 50–55°. Add slowly with stirring a BPHA soln. (0·2 g in 8–10 ml of ethanol) as required. Add dilute ammonia soln. to adjust the pH to 6·7. Stir the precipitate thoroughly at intervals. Allow it to settle, filter through a No. 4 sintered glass crucible, wash with distilled water, dry at 110–15° for 1 hr, cool and weigh.

Weight of lanthanum complex × 0·1791 = weight of lanthanum.

Separation of Iron, Aluminium, Thorium, Scandium, Gallium and Indium from Lanthanum

For the prior precipitation of iron, aluminium, thorium, scandium, gallium and indium, follow the procedures described for their determinations. Precipitate *iron* at pH 3·5 and *aluminium* at pH 4·0–4·5. Filter and wash the precipitates with an aqueous 0·1% BPHA soln. at pH 4·0–4·5 and determine lanthanum in the filtrate after adjusting to pH 6·7.

Before the precipitation of *thorium*, adjust the pH to 4·5–5·0 with a 10% ammonium acetate soln. Digest the precipitate at 60–70° on a water bath, filter, wash the precipitate with a hot aqueous 0·1% BPHA soln. at pH 5 and determine lanthanum in the filtrate. For *scandium*, adjust to pH 5·4 with ammonium acetate and dilute acetic acid solns. Precipitate the element with a boiling aqueous BPHA soln., filter, wash and determine lanthanum in the filtrate. Precipitate *gallium* and *indium* at pH 2·0 and 4·5, respectively, and determine lanthanum in the filtrate.

Separation of Uranium from Lanthanum

Dilute the test soln. to 150 ml and add to it an ethanolic BPHA soln. (8–10 ml) sufficient to precipitate both uranium and lanthanum. Add dilute ammonia soln. to adjust the pH to 5·2. Keep it below 50° on a water bath. Stir occasionally until red crystals of the uranium complex separate and the solution becomes colourless. Filter, wash the precipitate and from the combined filtrate and washings precipitate lanthanum as described above, after adjusting the pH to 6·7.

Separation of Lanthanum from Cerium(IV)

To 150 ml of test soln. add an ethanolic BPHA soln. (8–10 ml), enough to precipitate both ions. Adjust to pH 5. Keep below 50° on a water bath for 2 hr with occasional stirring till the precipitate becomes more or less granular. Filter, wash several times with a cold aqueous 0·1% soln. of the reagent at pH 5. From the combined filtrate and washings precipitate lanthanum by raising the pH to 6·7.

Note. The presence of acetate and a large amount of chloride interferes with the separation of cerium.

Separation of Lanthanum from Zinc, Copper, Nickel and Iron(II)

To 150 ml of test soln. add a little sodium sulphite to reduce any iron(III) present and add potassium cyanide in sufficient quantity to mask the interfering ions. Adjust the pH to 6·7 by the addition of dilute hydrochloric acid and determine lanthanum according to procedure described above.

Note. Scandium precipitates quantitatively only above pH 5·2.

DETERMINATION OF MERCURY IN THE PRESENCE OF MANY OTHER IONS; SEPARATION FROM MOLYBDENUM
(Das and Shome[19])

BPHA forms with mercury(II) a bright yellow complex, $Hg(C_{13}H_{10}O_2N)_2$, which is only slightly soluble in ethanol, ether, chloroform, carbon tetrachloride and ethyl acetate, but more soluble in dioxan, benzene and tetrahydrofuran. It dissolves easily in mineral acids and decomposes at 119–20°.

The pH range for the quantitative precipitation of mercury(II) is 3 to 6 and the amount of BPHA required is about 2·5 to 3·0 times the stoichiometric amount. The precipitate can be dried at 105° and weighed. Mercury(II) thus can be determined in presence of zinc, thallium(I), bismuth, antimony, cobalt, nickel, cadmium, lead, silver, tin(IV), indium, arsenate and tungstate with the use of suitable masking agents such as fluoride for indium, oxalate for tin(IV), tartrate for arsenic, antimony, bismuth and tungstate, and citrate for cobalt, nickel, cadmium, lead and silver. Molybdate, if present, should first be removed from a 1 N acidic solution and mercury determined in the filtrate. Chloride, cyanide and EDTA interfere with the precipitation of mercury.

Determination of Mercury(II)

Dilute the soln. (containing 10–16 mg of mercury) to 125–50 ml. Add a 10% sodium acetate soln. to adjust the pH to 4. Heat to 50–60°. Add dropwise with stirring a 1% ethanolic BPHA soln. (15–20 ml). Digest at 50–60° on a water bath (Note 1) for 30 min with frequent stirring. Filter through a No. 3 sintered glass crucible, wash with hot water followed by 20% ethanol (15–20 ml). Dry at 105° (Note 2) for 1 hr and weigh.

Weight of mercury complex × 0·3209 = weight of mercury.

Separation from Other Ions

For the determination in the presence of zinc, follow the above method but adjust the pH of the soln. to 3. Thallium(I) also does not interfere.

If bismuth, antimony, arsenate or tungstate is present, add in excess of sodium potassium tartrate. If cadmium, nickel, cobalt, lead or silver is present, add an excess of sodium citrate; for tin(IV) and indium, add sodium oxalate and sodium fluoride, respectively. Dilute each soln. to 150 ml, add a 10% sodium acetate soln. to keep the pH at 3·5–4·0, heat, and precipitate and determine mercury.

Separation of Mercury from Molybdenum

Add to the test soln. enough 2 N nitric acid to raise the acidity of the 100 ml of soln. to 1 N. Precipitate molybdenum in accordance with the procedure suggested (p. 69). Filter, and wash the precipitate with a warm (60°) 0·02 N nitric acid.

Dilute the filtrate to 150 ml, adjust to pH 4 with sodium acetate, and precipitate and determine mercury.

Notes

1. The temperature of the soln. and of the wash water must not be more than 60°.
2. The temperature should not exceed 105°.

SEPARATION OF BISMUTH FROM MOLYBDENUM, VANADIUM, TUNGSTEN, TITANIUM AND IRON; THE DETERMINATION OF BISMUTH IN THE PRESENCE OF OTHER IONS
(Das and Shome[20])

BPHA precipitates bismuth quantitatively from a tartrate solution at pH 6·0–6·8, when the supernatant solution contains about 0·05% of the reagent in excess. The creamy-white, granular complex so precipitated has the composition $Bi(C_{13}H_{10}O_2N)_3$. It is insoluble in water and sparingly so in ether, acetone, benzene, carbon tetrachloride, ethanol, acetic acid and ethyl acetate, but freely soluble in chloroform. It decomposes in moderately concentrated mineral acids and the same tendency is observed when digested in water above 65°, or when the pH of the solution is taken above 7·6. The complex melts with decomposition at 172 ± 1° but it maintains its composition at 110–15°, so it can be weighed as such after drying.

Cyanide, citrate, tartrate, oxalate, fluoride, phosphate and ascorbic acid do not interfere with the determination of bismuth. EDTA and thioglycollic acid, however, mask bismuth. Slightly low results for bismuth are obtained when tungstate, vanadate, titanium(IV) and iron(III) are removed, at a pH less than 4·5, before its determination in the filtrate. But when molybdenum(VI) is the interfering ion, such prior removal gives quite accurate results for bismuth.

Determination of Bismuth

Add sodium potassium tartrate (2·0–2·5 g) to a nitrate soln. containing bismuth (6·47–19·41 mg), dilute to 200 ml and heat to 50–55°. Add dropwise a BPHA soln. (0·2–0·4 g in 10–15 ml of ethanol), followed by dilute ammonia soln. dropwise with stirring to raise the pH to 6·0–6·8. Digest the bismuth precipitate for 30–45 min on a hot-water bath. The temperature, of the soln. must not be more than 65°. When the precipitate has settled and the soln. is clear, filter through a No. 3 sintered glass crucible, wash the precipitate with warm water, dry at 110–15° and weigh.

Weight of bismuth complex × 0·2471 = weight of bismuth.

Separation of Bismuth from Molybdate, Vanadate, Tungstate, Titanium(IV) or Iron(III)

To the soln. containing bismuth (6·47 mg) and either molybdenum(VI) (13·5 mg), tungsten(VI) (21·0 mg), vanadium(V) (14·0 mg), titanium(IV) (12·0 mg) or iron(III) (15·0 mg), add sodium potassium tartrate (2·0–2·5 g) and then precipitate with BPHA the appropriate ion at its optimum pH of precipitation. Filter, wash the precipitate with a tartrate soln. maintained at the pH of precipitation.

Concentrate the filtrate and washings to 200 ml. Precipitate bismuth at pH 6·0–6·4 with more BPHA in ethanol. Wash and dry and weigh the precipitate as above.

Separation of Bismuth from Lead, Arsenic(V), Antimony(III), Thorium, Uranium(VI), Lanthanum or Beryllium

Add to the test soln. sodium potassium tartrate (2 g) and sufficient sodium citrate and dilute to 200 ml. Add BPHA soln., adjust the pH to 6 with dilute ammonia soln., digest the precipitate and filter. Wash the precipitate first with a soln. at pH 6 containing tartrate and citrate and then with warm water. Dry and weigh the precipitate.

Separation of Bismuth from Tin(IV) or Aluminium

Use oxalate in place of citrate and follow the procedure given immediately above.

Separation of Bismuth from Copper, Cobalt, Nickel, Zinc, Manganese, Cadmium, Mercury(II) or Palladium

Add sodium potassium tartrate (2 g) and sufficient potassium cyanide, dilute to 200 ml and adjust the pH to 6·0–6·4. Precipitate and determine bismuth as described above.

DETERMINATION OF TITANIUM
(Shome[1])

Titanium has been determined by precipitation with BPHA from an acidic solution maintained below 25° and by ignition of the precipitate to the oxide as the weighing form. Tartrate and aluminium do not interfere.

Procedure

To the soln. (containing 0·1 g of TiO_2) add 6 N ammonia soln. to neutralize the free acid. Acidify with conc. hydrochloric acid (5 ml) and dilute to 400 ml. Cool, and precipitate titanium by adding dropwise with stirring, a 10% ethanolic BPHA soln. (about twice the stoichiometric quantity). Always keep the temperature of the soln. below 25°. Allow it to stand for 45 min with frequent stirring. Filter through a paper, wash the precipitate with water containing 10 ml of the reagent soln. per litre and conc. hydrochloric acid (3 ml), ignite, and weigh as TiO_2.

Weight of titanium oxide (TiO_2) × 0·5995 = weight of titanium.

SEPARATION OF TITANIUM FROM MOLYBDENUM AND DETERMINATION IN THE PRESENCE OF THORIUM, CERIUM, VANADIUM, IRON, ZIRCONIUM AND COPPER AND IN BAUXITE
(Kaimal and Shome[21])

The method of Shome[1] has been extended to cover the separation of titanium from many other elements by pH control and masking with complexing agents like EDTA and hydrogen peroxide.

To separate titanium from molybdenum, the latter is first precipitated and titanium is determined in the filtrate after demasking.

Determination of Titanium in the presence of Iron, Vanadium, Zirconium or Copper

Neutralize the titanium sulphate soln. (containing 6–14 mg of titanium) diluted to 250 ml and containing any one of the foreign ions with 6 N ammonia soln. till it smells faintly ammoniacal or till a slight turbidity appears. Add conc. hydrochloric acid (4 ml), followed by a 10% disodium EDTA soln. If vanadium is present as vanadate, boil the solution for 2 min to reduce the vanadium(V) to vanadium(IV). Cool, and add 2% BPHA in ethanol (10–15 ml) with stirring. Allow the precipitate to stand for 45 min (Note). Stir occasionally, filter, wash the precipitate with 0·033 N hydrochloric acid containing 1 ml of reagent soln. per 100 ml of acid, ignite and weigh as TiO_2.

Separation of Titanium from Molybdenum

Acidify about 250 ml of soln. containing titanium (6–14 mg) and molybdate (5–15 mg of molybdenum) with conc. hydrochloric acid (4 ml) after neutralization with 6 N ammonia soln. as described above. Add in succession, with stirring, 20 vol. hydrogen peroxide (2 ml), 10% disodium EDTA soln. (10 ml) and 2% BPHA soln. (10 ml) to precipitate molybdenum. Filter, and wash the precipitated molybdenum complex with 0·02 N hydrochloric acid followed by distilled water (10 ml). To the combined filtrate and washings, add sodium sulphite (1–2 g) and boil for 10 min to remove hydrogen peroxide. Cool, and add a soln. containing calcium chloride (1·5–2·0 g). Filter off any separated organic matter, wash the filter with a little water, adjust the total filtrate to pH 1–2 and determine titanium according to the method suggested above.

Separation of Titanium from Thorium or Cerium

To the test soln. add (if required) hydroxylammonium chloride, and heat to reduce all cerium(IV) to cerium(III). Cool, neutralize with 6 N ammonia soln., add conc. hydrochloric acid (4 ml), dilute to 250 ml and determine titanium as described above.

Determination of Titanium in Bauxite

Mix thoroughly a weighed quantity of bauxite (1·5–2·0 g), containing about 5% of titanium, with conc. sulphuric acid (10 ml) and distilled water (50 ml). Digest, heat to copious fumes, extract twice with 3–4 N hydrochloric acid, filter off the residual silica, wash thoroughly and dilute the combined filtrate and washings to exactly 250 ml. Take an aliquot (25–50 ml) of this soln., dilute to 300 ml, add 10% EDTA soln. (10–15 ml) and adjust the pH to 1–2. Precipitate and estimate titanium as described above.

Note. Sulphate, even when present in small quantities, precipitates the zirconium–EDTA complex slowly on long standing.

SEPARATION AND DETERMINATION OF TITANIUM BY WEIGHING THE COMPLEX
(Kaimal and Shome[22])

Recently, a method for the determination of titanium by direct weighing as $TiO(C_{13}H_{10}O_2N)_2$, after precipitation from 0·02–0·50 N sulphuric acid at 65–75°, has been proposed. Under these conditions, thorium, cerium(III), uranium(VI), chromium(III), aluminium, zinc, manganese, nickel and cobalt remain in solution; EDTA is needed to mask iron(III), vanadium(IV) and copper. Tartaric acid is added when titanium is determined in the presence of phosphate.

The titanium precipitate tends to be contaminated with the free acid of EDTA; hence it is preferable to use magnesium–EDTA soln. in place of disodium EDTA alone.

For the complete precipitation of titanium as $TiO(C_{13}H_{10}O_2N)_2$, at least thrice the stoichiometric quantity of BPHA is required. The pale-yellow complex so obtained is soluble in chloroform, ether, benzene and 50% ethanol, is slightly soluble in water above 80°, but is insoluble in 15% ethanol and is decomposed by strong mineral acids and by dry heating to 167–8°.

Determination of Titanium by Direct Weighing of the Complex

Neutralize the soln. (containing 6–24 mg of titanium) with 6 N ammonia soln. and add enough 2 N sulphuric acid to maintain the acidity at 0·02–0·5 N on dilution to 200 ml. Dilute to 200 ml, heat to 65–75°, add dropwise with stirring a 3% soln. of BPHA in ethanol until the precipitation is complete. The amount of BPHA added must be at least 3 times the stoichiometric quantity or about 27 times the amount of titanium present. Heat on a water bath at 65–75° for 30 min more. Filter through a sintered glass crucible, wash the precipitate with water at 40–50°, dry at 110° and weigh.

Weight of titanium complex × 0·09808 = weight of titanium.

Separation of Titanium from Uranium(VI), Thorium, Cerium(III), Chromium(III), Aluminium, Cobalt, Zinc, Nickel and Manganese by pH Adjustment

Dilute the test soln. prepared as above to 200 ml with distilled water. Adjust the acidity to 0·5 N with sulphuric acid, and precipitate and determine titanium according to the above procedure.

If cerium(IV) is present, add hydroxylammonium chloride to reduce it and then proceed with the determination of titanium as described above.

Separation of Titanium from Iron(III), Vanadium(IV) or Copper

To the test soln. add an EDTA–magnesium soln. (20 ml of 0·25 M disodium EDTA and 5 ml of 1 M magnesium chloride solns.) or as required (Note 1), dilute to 200 ml, adjust the acidity, heat and determine titanium (Note 2).

Vanadate must be reduced to vanadium(IV) for complexation with EDTA. Thus to separate titanium from vanadium(V) (15–30 mg), add sodium sulphite (0·5 g) and 18 N sulphuric acid (2 ml), boil, cool and add EDTA–magnesium soln. (25 ml). Dilute to 200 ml, adjust the acid concentration, heat, and determine titanium (Note 3).

For the separation of titanium from phosphate (100–150 mg), add tartaric acid (2 g) and then follow the procedure for titanium determination.

Notes

1. Because disodium EDTA is not very soluble in water, it is advisable to use a 0·25 M soln.
2. Owing to the low solubility of the zirconium–EDTA–sulphate complex, titanium cannot be separated from zirconium under the conditions of precipitation from sulphuric acid.
3. Excess of sulphur dioxide should be boiled off before the precipitation of titanium, otherwise the digestion period for the complete precipitation of titanium must be prolonged.

SEPARATION OF TITANIUM FROM MOLYBDATE AND TUNGSTATE USING CUPFERRON AND DETERMINATION OF TANTALUM IN THE PRESENCE OF NIOBIUM, TITANIUM AND ZIRCONIUM

(Moshier and Schwarberg[23])

From a sulphuric acid solution of pH $1·0 \pm 0·1$ containing a small amount of hydrofluoric acid, not more than 50 mg of tantalum pentoxide can be precipitated by adding a hot aqueous solution of BPHA. The pH must not exceed 1·2 and the solution must not contain a large amount of oxalate or chloride.

As the tantalum compound is appreciably soluble in solutions above 27°, cooling in water at room temperature for a minimal period of 2 hr is required for quantitative precipitation. On the other hand, as the time of standing increases, so does the coprecipitation of niobium. The tantalum precipitate is ignited to the oxide as the weighing form.

For samples containing less than 10 mg of niobium pentoxide, reprecipitation is necessary. When the niobium pentoxide content is more than 10 mg, a second reprecipitation is required to reduce the impurity in the recovered tantalum oxide to less than 0·1 mg.

The results for the determination of 0·1 g of tantalum oxide are low by 2 mg or more. In the development of the method, and in the determination of the extent of contamination of the precipitates, niobium-95 and tantalum-182 tracers have been used. Interference due to zirconium and titanium is negligible compared to that found with niobium. Molybdate and tungstate, if present, need prior removal by washing the mixed niobium, tantalum, titanium and zirconium cupferrates with ammonia. Preliminary experiments indicate that tantalum can probably be separated from iron(II and III), nickel, cobalt, aluminium, chromium(III), tin(IV) and vanadium(IV).

Determination of Tantalum

To the weighed sample (0·1 g or less) in a platinum crucible, add conc. hydrofluoric acid (5 ml), cover the crucible and fill the cover with water to minimize the loss of acid during digestion. Heat at a low temperature on a hot plate for 15–30 min to dissolve the sample completely. Cool the crucible, rinse the inside surface of the cover with water and collect the rinsings in the crucible. Add from a graduated pipette 50% (v/v) sulphuric acid (3 ml) to give the required pH of 1·0 ± 0·1 to the total 300 ml of soln. from which precipitation occurs. Heat the crucible on the hot plate to volatilize hydrofluoric acid and continue heating till fumes of sulphuric acid are evolved and the product gives the appearance of a gel. Cool the crucible and treat the content with 4 M hydrofluoric acid (2 ml) to give a clear soln. Transfer completely to a 400-ml polyethylene beaker containing water (185 ml) and 4 M hydrofluoric acid (2 ml). Police the crucible with two 0·5-ml portions of 4 M hydrofluoric acid, rinse with water and add the washings to the main soln. in the beaker, which now contains a total of 5 ml of 4 M hydrofluoric acid.

Add with vigorous stirring a hot BPHA soln. (0·4 g in 100 ml of boiling water) to precipitate tantalum. A white opalescence appears. Cover the beaker and keep in a bath of tap water for 2·5 hr. Filter through a Whatman No. 42 filter paper by gravity, or by gentle suction with the use of a platinum filter cone. Wash the precipitate 8 times with a soln. containing 0·6 g of the reagent in 1 litre of water. This includes four to five policings for the complete recovery of the precipitate. Char, and ignite at 900° for 2 hr in a crucible and weigh as tantalum pentoxide.

Weight of tantalum oxide (Ta_2O_5) × 0·8189 = weight of tantalum.

Preliminary Removal of Molybdenum and Tungsten

Dissolve the sample in hydrofluoric acid as suggested above and follow the procedure up to the appearance of the gel. Add 4 M hydrofluoric acid (1 ml) to dissolve the gel and transfer the entire contents to a beaker holding 140 ml of water containing 4 M hydrofluoric acid (0·5 ml). Police the crucible twice with 0·5-ml portions of 4 M hydrofluoric acid, rinse with water and add the rinsings to the main soln. Adjust the pH to 3–4 by adding ammonium acetate (5 g) and ammonia soln., as required. Add some filter paper pulp to aid the filtration and washing of the precipitate, and then add a cold 0·2 M cupferron soln. (50 ml or more) with vigorous stirring to precipitate tantalum and also titanium, zirconium and niobium. Allow the mixture to stand in an ice bath for 2 hr to ensure complete precipitation. Niobium, particularly, precipitates more slowly under these conditions. Filter through a Whatman No. 40 filter paper, wash 8 times with an ice cold 0·02 M cupferron soln. and then 5 times with a 5% ammonia soln. to dissolve out any coprecipitated molybdenum and tungsten. Char the precipitate and ignite at 900° for 2 hr.

Determine tantalum, after dissolution of the ignited oxides, according to the procedure given above.

DETERMINATION OF NIOBIUM AND TANTALUM AND THEIR SEPARATION FROM EACH OTHER. SEPARATION OF NIOBIUM FROM VARIOUS IONS
(Majumdar and Mukherjee[24,25])

Niobium and tantalum are quantitatively precipitated by BPHA, from a tartrate soln. 2 N in sulphuric acid. The pH range for the complete precipitation of niobium extends to 6·5, while that for tantalum extends to pH 1·5. A higher pH tends to keep tantalum in solution. At pH 3·5–6·5 only niobium is completely precipitated; tantalum remains in solution, from which, however, it can be precipitated by acidification. With niobium : tantalum ratios of 1 : 16 to 100 : 1, a single precipitation is sufficient for the complete separation of niobium and tantalum from each other. For higher ratios even up to 1 : 100, only a second precipitation is necessary for the separation of niobium. With EDTA and tartaric acid as masking agents, niobium is determined at pH 3·5–6·5 in the presence of other ions such as of iron(III), uranium(VI), beryllium, thorium, chromium(III), aluminium, lanthanides, cerium(III and IV), copper, cadmium, bismuth, lead, mercury(II), arsenic(III), antimony(III), tin(IV), zinc, manganese, nickel, cobalt, calcium, strontium, barium, magnesium, phosphate, tungstate, chromate and arsenate. Titanium, zirconium, molybdate and vanadate must be absent.

The precipitate of tantalum is less granular than that of niobium but is easily filtered and washed. Both complexes are ignited to oxides as the weighing forms.

Determination of Niobium or Tantalum

Fuse a weighed sample of the oxide in a silica crucible with 10 times its weight of potassium hydrogen sulphate and extract the melt with a 5% soln. of tartaric acid. Add dilute sulphuric acid or a 20% ammonium acetate soln. to adjust the pH to the desired value, ($\not>$ 1·5 for tantalum and $\not>$ 6·5 for niobium). To raise the pH above 5·9, add a few drops of dilute ammonia soln. after the ammonium acetate. Dilute the solution to 200 ml, heat and add with stirring a 10% soln. of BPHA in ethanol until no more white turbidity appears. Keep the mixture on a boiling water bath for 45 min and stir occasionally to coagulate the precipitate. Filter through a quantitative filter paper, wash the precipitate until free from sulphate with boiling water containing 0·1 g of BPHA per 100 ml, dry, ignite and weigh as the pentoxide.

Weight of tantalum oxide (Ta_2O_5) × 0·8189 = weight of tantalum.
Weight of niobium oxide (Nb_2O_5) × 0·6990 = weight of niobium.

Separation of Niobium and Tantalum

Fuse the mixed oxides with potassium hydrogen sulphate and dissolve the melt in a 5% tartaric acid soln. as above. Add enough tartaric acid so that when the solution is finally diluted to 350 ml, the strength of tartaric acid is 5%. Dilute to 350 ml. Add the required quantities of 20% ammonium acetate soln. and dilute ammonia soln., adjust the pH of the soln. to 3·5-6·5. A pH of 6 is preferable to avoid contamination if tantalum is present in large excess.

Heat the soln. to boiling, precipitate niobium, filter, wash the precipitate with the hot wash liquid (0·1% aqueous BPHA soln.) and determine niobium as the oxide as described above.

To reprecipitate niobium, ignite the precipitate and fuse it with potassium hydrogen sulphate in a silica crucible. Extract with a 5% tartaric acid soln. and precipitate, filter, wash and determine niobium as above. Recover tantalum from the filtrate by acidifying the combined filtrate and washings with dilute sulphuric acid to pH 1. Heat the mixture to boiling and keep it on the boiling water bath for 45 min. Filter, wash, ignite the resulting precipitate and weigh as tantalum oxide.

When less than 5 mg of either element is present, use filter paper pulp to aid filtration.

In the presence of a large excess of tantalum or complexing agents such as tartaric acid and EDTA, increase the digestion period to about 90 min to allow the complete coagulation of the niobium precipitate.

Determination of Niobium in the Presence of other Ions

To the test soln. add in succession tartaric acid and EDTA, each in amounts 2 to 3 times the theoretical amount of the various ions. Adjust the pH to 3·5–6·5, and precipitate and determine niobium as described above.

SEPARATION OF NIOBIUM FROM OTHER IONS INCLUDING TANTALUM, AND THE DIRECT DETERMINATION OF NIOBIUM
(Majumdar and Mukherjee[26])

The buff-coloured niobium precipitate, formed on the addition of BPHA to a tartrate solution of niobium, maintained at a pH lower than 6·5, is insoluble in water, ethyl and amyl acetates, acetone, carbon tetrachloride and benzene, but is highly soluble in chloroform. The complex, when dried at 110° is of definite composition, $NbO(C_{13}H_{10}O_2N)_3$, and can be weighed as such, or can be ignited to oxide before weighing. The niobium complex, which may have been formed by the elimination of water, during the reaction of niobic acid with BPHA on a boiling water bath, is found from thermogravimetric analysis to be stable up to 229°.

The tantalum precipitate, on the other hand, is of indefinite composition, it loses weight as the temperature is increased and ultimately is converted, like the niobium complex, to the oxide, at 500°. The tantalum complex, however, is insoluble in all the solvents mentioned above.

Determination of Niobium

Precipitate niobium from a tartrate soln. at pH 3·5–6·5, according to the procedure suggested previously.[25] Digest the precipitate on a boiling-water bath for 1 hr, filter through a weighed, sintered glass crucible, wash thoroughly with hot water, dry at 110° and weigh.

Weight of niobium complex × 0·1783 = weight of niobium oxide (Nb_2O_5).
Weight of niobium complex × 0·1246 = weight of niobium.

Separation of Niobium from Tantalum and Other Ions; Determination of Niobium

Follow the procedures as described[25] for the separation (i) of niobium and tantalum from each other and (ii) of niobium from other ions in presence of EDTA and tartaric acid. Wash the niobium precipitate first with a hot 0·1% aqueous reagent soln. until free from sulphate and then with hot water. Dry at 110° and weigh. From the combined filtrate and washings of (i), determine tantalum after precipitation at pH 1 and ignition to oxide, as on p. 91.

DETERMINATION OF NIOBIUM AND TANTALUM IN PRESENCE OF VARIOUS IONS AND IN STEEL
(Mukherjee[27], Majumdar and Mukherjee[28])

Whereas niobium and tantalum are completely precipitated from 10% (v/v) sulphuric acid at room temperature by BPHA, zinc, manganese, nickel, cobalt, copper, cadmium, mercury(II), beryllium, magnesium, uranium(VI), chromium(III), iron(III), aluminium, cerium(III and IV), arsenic(III), thorium, phosphate, chromate and arsenate remain in solution. The ions which form insoluble sulphates and titanium, zirconium, molybdate, vanadate and tungstate interfere. The niobium and tantalum precipitates are ignited to their pentoxides as the final weighing forms, since the precipitates obtained from such a strongly acidic solution are not of any definite composition.

These conditions, which allow the quantitative separation of niobium and tantalum from many other ions, are equally effective for the determination of niobium and/or tantalum in steels.

Separation of Niobium or Tantalum from Various Ions

To the tartrate soln. of niobium or tantalum containing other ions and diluted to 250 ml, add enough conc. sulphuric acid to raise the acidity to 10% (v/v). Cool to room temperature and add a 10% ethanolic soln. of BPHA slowly with stirring to precipitate completely niobium or tantalum. Stir occasionally and keep the soln. at room temperature for 30 min. Filter, wash the precipitate with hot water, ignite and weigh as the oxide. Solutions more than 2 N in acid must not be heated, otherwise a gummy precipitate, difficult to filter, is likely to appear.

Analysis of Straight Nb 18/12 Stainless Steel Sample No. 261
(British Chemical Standard Containing Carbon, 0·083%, Silicon, 0·39%, Chromium 17·2%, Nickel 13·08%, Niobium and Tantalum, 0·71% and Manganese, 0·66%).

Take a weighed amount of steel (1–2 g) in a 600-ml beaker and add (1 + 3) sulphuric acid (60 ml). Heat to dissolve and evaporate on a hot plate until sulphuric acid fumes are evolved. Cool to room temperature and add to it water (300 ml) containing tartaric acid (10 g). Heat again to dissolve the iron salts. Ignore any particle remaining undissolved. Add more conc. sulphuric aid (15 ml), bring the acid concentration to 10% (v/v), precipitate niobium by adding with stirring a 10% ethanolic BPHA soln. (5 ml). Keep it at room temperature for 30 min more, filter through a quantitative filter paper, wash the precipitate with hot water and ignite in a silica crucible. Fuse the ignited product with 10 times its weight of potassium hydrogen sulphate, dissolve the clear melt in a 5% tartaric acid soln. filter, and wash the residue, if any, with hot dilute sulphuric acid. Adjust the acidity of the combined filtrate and washings to 10% (v/v) sulphuric acid, precipitate niobium as described, filter, wash the precipitate, ignite and weigh.

Analysis of Nb-Mo 18/12 Stainless Steel Sample No. 246
(British Chemical Standard Containing Carbon, 0·062%, Niobium, 0·82%, Molybdenum, 2·89%, Chromium, 18·8%, Nickel, 12·1%, Tungsten, 0·22% and Copper 0·13%)

Follow the procedure given above for the dissolution of the steel sample as far as bringing the acidity to 10% (v/v) in sulphuric acid. Cool in an ice bath. Add a suspension of ashless paper pulp (few ml) and a 6% aqueous soln. of cupferron (5 ml), with stirring, so that all the niobium, molybdenum and tungsten, and some iron, precipitate. Wait for about 15 min for the precipitates to settle. Filter, wash with cold 5% ammonia soln. (100 ml) to dissolve out practically all the molybdenum and tungsten, and ignite the precipitate in a silica crucible.

Fuse the ignited mass with 10 times its weight of potassium hydrogen sulphate, dissolve the melt in a 5% tartaric acid soln., filter off the residual silica and wash with hot $(1+4)$ sulphuric acid. Add to the combined filtrate and washings about 2 g disodium EDTA, adjust the pH to 6 by adding a 20% ammonium acetate soln. and dilute ammonia soln., heat to boiling and precipitate niobium by adding a 10% soln. of BPHA in ethanol. Digest for 1 hr on a boiling-water bath. Remove from the water bath, filter, wash the precipitate with hot water, ignite and weigh as the pentoxide or dry at 110° and weigh as the complex.

SEPARATION AND DETERMINATION OF NIOBIUM, TANTALUM AND TITANIUM
(Langmyhr and Hongslo[29])

This method proposes the use of EDTA and hydrogen peroxide to keep titanium in solution, while niobium and tantalum are precipitated together from 1 N sulphuric acid and then separated from each other, according to the methods suggested by Majumdar and Mukherjee.(24–26) For the separation of niobium and tantalum from each other, the most suitable pH for the precipitation of niobium is suggested as 4·5–5·0, since a higher pH (5–6) may cause less satisfactory coagulation of the precipitate and effect the precipitation of the hydrous oxide.

The amount of BPHA added for every precipitation of titanium, niobium and tantalum is 8–10 times the combined weights of the metals present. During the separation of niobium and tantalum from titanium, about 10% of the titanium coprecipitates; hence a second precipitation is preferred.

At a pH less than 1, niobium, tantalum and titanium are quantitatively separated by a double precipitation with BPHA, several other elements being kept in solution by complex formation with EDTA and tartaric acid, as recommended previously.(24,25)

During the preliminary separation of niobium, tantalum and titanium from the other elements, small amounts of iron and other constituents tend to coprecipitate. They remain dissolved, however, in subsequent separations, and the filtrate from the titanium precipitation is retained for their determinations.

Determination of Niobium, Tantalum and Titanium

Fuse a weighed amount (0·2–0·3 g) of the oxides of titanium, niobium and tantalum with potassium pyrosulphate (2–3 g) and dissolve the melt in a 30% tartaric acid soln. (30–40 ml containing 2 ml of conc. sulphuric acid). Transfer the resulting soln. to a 800 ml beaker and dilute to 300 ml with a 2% tartaric acid soln. Add sufficient (1 + 1) sulphuric acid to make the soln. 1 N in acid. Heat to 60° on a boiling-water bath, remove from the bath and add in succession EDTA (1 g), 30% hydrogen peroxide soln. (1 ml) and, dropwise with stirring, a 10% soln. of BPHA in ethanol (30 ml). Keep it at 60° on the water bath for 45 min till the precipitate coagulates. Add macerated filter paper, cool the soln. first to room temperature and then in a refrigerator to 5°. Filter the yellow precipitate through paper; wash it until free from sulphate with a hot aqueous soln. containing BPHA (1 g), EDTA (1 g) and 30% hydrogen peroxide soln. (1 ml) per litre. Retain the filtrate and washings for the titanium determination. Dry the precipitate in the filter paper in a porcelain crucible, char and ignite at 600° in an electric furnace.

Fuse the ignited oxides containing mainly niobium and tantalum with potassium pyrosulphate, dissolve, and precipitate niobium and tantalum with BPHA as before but adding only 0·5 g of EDTA and 0·5 ml of the 30% hydrogen peroxide soln. to keep the remaining titanium in soln. Retain the filtrate and washings.

Fuse the ignited oxides (containing niobium, tantalum and a small quantity of titanium) again with potassium pyrosulphate. Dissolve the melt as before in the 30% tartaric acid soln. (30–40 ml containing 2 ml of conc. sulphuric acid). Transfer and dilute to 300 ml in a beaker with the 2% tartaric acid soln. Add, with stirring, a 10% ammonium acetate soln. (20 ml) and dilute ammonia soln. until the pH is 4·5–5·0. Heat to 90° on the water bath and add slowly with stirring a 10% soln. of BPHA in ethanol (30 ml). Keep for 30–45 min on the water bath with occasional stirring. Add macerated filter paper. Remove the beaker from the water bath, cool to room temperature, filter and collect the precipitate on the filter paper. Wash the precipitate with a hot 0·1% soln. of BPHA in water, dry, char, ignite in a porcelain crucible at 900° to niobium pentoxide and weigh.

To the combined filtrate and washings (which contain tantalum and a very small quantity of titanium) add EDTA (0·5 g) and 30% hydrogen peroxide soln. (0·5 ml). Heat to 60° on the water bath. Lower the pH to 1 by adding (1 + 1) sulphuric acid. Add a 10% BPHA soln. in ethanol (20 ml) slowly with stirring. Keep at 60° on the water bath for 1 hr. Add macerated filter paper, treat the precipitate in the same way as described above for the niobium precipitate and weigh as tantalum pentoxide.

Evaporate the combined filtrate and washings from the tantalum precipitation and all other washings containing titanium, on a boiling water bath. Add fuming nitric acid to decompose the organic matter and heat until copious fumes of sulphuric acid appear. Dilute the clear soln. with water so that it is 5% in sulphuric acid. Heat on a boiling water bath. Add macerated filter paper and add, with stirring, a 10% BPHA soln. in ethanol to precipitate titanium. Leave it on the water bath for 45–60 min, stirring occasionally. Cool to room temperature and then in the refrigerator to 5°. Filter, wash, ignite as described for the niobium precipitate and weigh as titanium dioxide.

SEPARATION AND DETERMINATION OF NIOBIUM AND TANTALUM AND THE SPECTROCHEMICAL EXAMINATION OF THE PRECIPITATES

(Majumdar and Pal[30] and Pal[31])

This investigation confirms the previous findings of Majumdar and Mukherjee[25,26] that within a coagulation period of 30–35 min, niobium and tantalum can be separated from each other from a tartrate solution at pH 3·5–6·5 with highly satisfactory results.

Tantalum, which remains in solution, is quantitatively precipitated from the filtrate at pH 1. Moreover, there is no possibility of the precipitation of hydrous oxides in the presence of complexing agents such as tartrate, even when the solution containing niobium or tantalum is made ammoniacal.

Spectrochemical examination of the niobium complex obtained after precipitation at pH 3·5–6·5 in the presence of tantalum, and of the tantalum precipitates obtained from the filtrate at pH 1, shows the niobium precipitate to be practically free from tantalum, and the tantalum precipitate to contain only traces of niobium.

DETERMINATION OF GERMANIUM
(Alimarin, Sokolova and Smolina[32])

The precipitation of germanium by BPHA is incomplete from hydrochloric or sulphuric acid, but from a hot perchloric acid solution ($\geqslant 2$ N) germanium precipitates quantitatively as $Ge(C_{13}H_{10}O_2N)_4$ on the addition of an ethanolic solution containing at least 1·5 times the stoichiometric amount of BPHA. However, weighing the precipitate after drying at 110° does not give satisfactory results. The precipitate decomposes at a temperature above 320° and is soluble in alkaline solutions as well as in organic solvents.

Procedure

To a 10-ml soln. (containing 5–20 mg of germanium) add perchloric acid to adjust the acidity to 2 N. Heat on a boiling-water bath and add a 5% ethanolic BPHA soln. (8–10 ml). Heat for another 10–15 min with slow stirring. Allow the soln. to cool, filter through a filter paper, wash the precipitate with water until it is free from perchlorate, ignite to the oxide at 800–900° and weigh.

Weight of germanium oxide (GeO_2) × 0·6941 = weight of germanium.

DETERMINATION OF MAGNESIUM AND SEPARATION FROM BERYLLIUM, COPPER, NICKEL, COBALT, ZINC, IRON, ALUMINIUM AND THORIUM
(Cardwell and Magee[33])

From an ammoniacal solution, adjusted to pH 8–9, magnesium precipitates quantitatively as $Mg(C_{13}H_{10}NO_2)_2$ on the addition of a

1·5–2·0-fold excess of BPHA in ethanol. The optimum temperature for the reaction is 55–60°. Prolonged digestion of the precipitate should be avoided as the precipitate then absorbs a yellow colour produced by the decomposition of the reagent. The precipitate is ignited to the oxide as the final weighing form.

Procedure

Dilute the soln. (containing 2–20 mg of magnesium) to 150 ml, add ammonium chloride (1 g) and heat to 60°. Adjust the pH to 8–9 by adding conc. ammonia soln. and add in quick succession with stirring an ethanolic BPHA soln. (10 ml, containing 1·5 times the stoichiometric amount of BPHA). Allow the precipitate to settle for 10–15 min. Filter through Whatman No. 42 filter paper, wash the precipitate, ignite to the oxide and weigh.

Weight of magnesium oxide (MgO) $\times$ 0·6032 = weight of magnesium.

Separation from Other Elements

First precipitate the elements which form insoluble compounds at a lower pH, leaving magnesium in the filtrate to be determined later according to the above procedure.

Procedure

For the separation of beryllium, precipitate the element at pH 5, adjusted by dilute ammonia soln., filter and wash the beryllium precipitate with an 0·1 % ethanolic soln. of BPHA at pH 5. Concentrate the filtrate and determine magnesium.

In the same way, precipitate iron(III) at pH 3·5 and aluminium at pH 4·5. Filter, wash the precipitates with an 0·1 % ethanolic BPHA soln. at pH 4·0–4·5 and determine magnesium in the filtrate.

To separate thorium, adjust the pH to 4·5–5·0 with a 10 % ammonium acetate soln., add BPHA in ethanol, digest at 65°, filter, and wash the precipitate with 0·1 % BPHA in ethanol at pH 5, and determine magnesium in the filtrate.

For the removal of copper, nickel, cobalt and zinc, precipitate the copper at pH 4·0 and the other elements at pH 5·5, using a 10 % sodium acetate soln. for the adjustment of pH and hydroxylammonium chloride to keep cobalt in the bivalent state. Digest for 1 hr on a water bath, filter, wash the precipitates and determine magnesium in the filtrate as above.

As barium and strontium are not precipitated by BPHA, magnesium can be determined in their presence. Calcium, however, co-precipitates and hence its prior removal is necessary. Sulphate and phosphate must be absent.

2. Applications of N-Cinnamoylphenylhydroxylamine

DETERMINATION OF NIOBIUM AND TANTALUM AND THEIR SEPARATION FROM OTHER IONS
(Majumdar and Mukherjee[34])

N-Cinnamoylphenylhydroxylamine (CPHA) precipitates niobium and tantalum from oxalate or tartrate solutions. The acid range at which niobium is completely precipitated is from 10% (v/v) sulphuric acid to pH 7·5; that for tantalum is only 2·5–10·0% (v/v) sulphuric acid. At a lower acidity, the tantalum precipitate is colloidal. The precipitates of niobium and tantalum are ignited to the oxides as the final weighing form, because the sparingly soluble reagent is always associated with the precipitate. The reagent does not allow the separation of niobium and tantalum from each other, but allows their separation from other ions such as zinc, manganese, nickel, cobalt, magnesium, beryllium, copper, mercury(II), cadmium, arsenic(III), bismuth, aluminium, chromium(III), uranium(VI), thorium, cerium(IV), phosphate and arsenate present in 10% (v/v) sulphuric acid solution.

Determination of Niobium or Tantalum

Fuse a weighed quantity of the oxide in a silica crucible with 10 times its weight of potassium hydrogen sulphate and dissolve the melt in a 5% tartaric acid soln. or in a 2% ammonium oxalate soln. Dilute to 250 ml. To determine niobium, add sulphuric acid or dilute ammonia soln. so that the hydrogen ion concentration is between 10% (v/v) in sulphuric acid and pH 7·5. To determine tantalum, keep the acidity between 2·5 and 10·0% (v/v) in sulphuric acid. If the pH is >2, heat the soln. nearly to boiling, add a few ml of a 2·5% soln. of CPHA in ethanol with stirring (7·5 ml of the 2·5%CPHA soln. are required for every 10 mg of the pentoxide), keep the soln. on a boiling-water bath for 30–60 min for the complete coagulation of the precipitate and stir occasionally. Remove from the water bath and cool to room temperature. Filter, wash the precipitate with warm water, dry, char, ignite and weigh as the pentoxide.

If the pH of the soln. is <2, carry out the precipitation at room temperature (25–30°) to prevent the formation of a gummy precipitate.

For the separation of niobium and/or tantalum from other ions, make the test soln. containing tartaric or oxalic acid, 10% (v/v) in sulphuric acid and follow the procedure as described above.

3. Applications of *N*-Benzoyl-*o*-tolylhydroxylamine

DETERMINATION OF NIOBIUM AND TANTALUM AND SEPARATION FROM EACH OTHER AND FROM OTHER IONS
(Pal[31] and Majumdar and Pal[35])

N-Benzoyl-*o*-tolylhydroxylamine precipitates niobium and tantalum from a tartrate solution. For the quantitative precipitation of niobium, the acid range is from 10% (v/v) sulphuric acid (sp. gr. 1·8) to pH 6·7, and for tantalum the range is from 10% (v/v) sulphuric acid to pH 1·8. The best pH range for the complete separation of niobium and tantalum from each other is between 5·0 and 6·5, provided the tartrate solution is cold, because tantalum does not precipitate from the cold solution at pH $\geqslant$ 5; if it is digested on a hot-water bath it precipitates even at pH 6. In the cold, the reagent affords, simply by one precipitation, a highly satisfactory separation of niobium from tantalum, even when the elements are present in the ratios of 1:17 to 21:1. Spectrochemical analysis shows the niobium precipitates to be free from tantalum.

Niobium and tantalum may be separated from several other ions such as chromium(III), beryllium, magnesium, cadmium, aluminium, lanthanum, cerium(III), iron(III), thorium, uranium(VI), cobalt, nickel, copper, mercury(II), zinc, phosphate, arsenate and vanadate by precipitation from 10 per cent (v/v) sulphuric acid. Tungstate, molybdate, titanium and zirconium interfere. The interference of tungstate and molybdate in the niobium determination is eliminated if the niobium is precipitated at pH 4·5–6·0, adjusted by ammonium acetate. Spectrochemical analysis shows the combined niobium and tantalum precipitates, to be free from any contaminating ions.

Both the niobium and tantalum precipitates are ignited to the pentoxides as the final weighing forms. The yellow niobium complex, though of definite composition, $(C_{14}H_{12}NO_2)_3NbO$, and stable for several hours at 110°, cannot be weighed directly because of its contamination with the reagent. The niobium complex is slightly soluble in hot water and is appreciably soluble in ethanol and chloroform. The white tantalum complex is of indefinite composition and loses weight as the temperature is raised.

Determination of Niobium or Tantalum

Fuse the weighed oxide with 10 times its weight of potassium hydrogen sulphate and dissolve the clear melt by extraction with a mixture of 5% (v/v) sulphuric acid (sp. gr. 1·8) and 5% (w/v) tartaric acid solns. Dilute the soln. to 100 ml with water. Add ammonium chloride (2–3 g). Adjust the pH either with a 20% (w/v) ammonium acetate soln. and dilute ammonia soln., or with dilute (1:1) sulphuric acid. Maintain the acidity between 10% (v/v) sulphuric acid and pH 6·7 for niobium and between 10% (v/v) sulphuric acid and pH 1·8 for tantalum. Add slowly, with thorough stirring after each addition, a soln. of N-benzoyl-o-tolylhydroxylamine in the minimal amount of 1:1 aqueous ethanol till precipitation is complete. Allow the precipitate to settle for 1–2 hr, filter, wash until free of sulphate with a soln. prepared by mixing a soln. of the reagent (0·1 g in 2–3 ml of ethanol) with an ammonium chloride soln. (1–2 g in 100 ml), and then with a few ml of cold water. Dry the precipitate, ignite and weigh as the pentoxide.

Separation of Niobium from Tantalum

Fuse the mixed oxides with potassium hydrogen sulphate and dissolve as above. Dilute to 250 ml. Add ammonium chloride (2–3 g) and a 20% ammonium acetate soln. and dilute ammonia to maintain the pH between 5·0 and 6·5. Precipitate by adding reagent about 20 times in excess of the amount the total oxides present and determine niobium as above.

To the combined filtrate and washings, add with stirring a few ml more of reagent soln. and then (1 + 1) sulphuric acid to make the soln. about 10% (v/v) in sulphuric acid. Stir well. Allow the tantalum precipitate to settle, filter, wash, dry, ignite and weigh.

Separation of Niobium and Tantalum from Other Ions

Dilute to 250 ml the tartrate soln. of niobium and/or tantalum and other ions, containing an amount of tartaric acid equivalent to at least 2–3 times the total amount of metal ions present. Acidify the soln. to 10% (v/v) sulphuric acid. Precipitate and determine niobium and/or tantalum, according to the method described above. If iron(III) or vanadate is present in large excess, again fuse the oxides obtained by ignition of the precipitate with potassium hydrogen sulphate, dissolve in tartaric acid and reprecipitate niobium and/or tantalum from a 10% (v/v) sulphuric acid soln., as described above.

For the separation of niobium, and not tantalum, from molybdate and tungstate, follow the same procedure but keep the pH between 4·5 and 6·0.

4. Application of N-Salicylphenylhydroxylamine

DETERMINATION OF TITANIUM
(Ghosh and Bhattacharyya[36])

In the presence of EDTA at pH 6·6, titanium(IV) is completely precipitated by N-salicylphenylhydroxylamine as $(C_{13}H_{10}O_3N)_3TiOH$, which can be dried and weighed as such.

Procedure

Add a two-fold excess of disodium EDTA soln. (0·1 M) to the titanium (IV) soln. (0·0375 M) followed by about 3 times the stoichiometric amount of N-salicylphenylhydroxylamine in aqueous soln. (0·2 M, 5 g of reagent dissolved in the least quantity of dilute ammonia soln. and diluted to 100 ml). Adjust the pH to 6·6 by the dropwise addition of dilute ammonia soln. with stirring. Allow the precipitate to settle for 30 min. Filter through a sintered glass crucible, wash with water, dry at 95–100° and weigh.

Weight of titanium complex × 0·0639 = weight of titanium.

5. Application of N-Acetylsalicylphenylhydroxylamine

DETERMINATION OF TITANIUM
(Savariar and Joseph[37])

Titanium(IV) can be precipitated as $TiO(C_{15}H_{12}O_4N)_2$ from 1–2 N hydrochloric acid on the addition of 2–3 times the stoichiometric quantity of the reagent. Because the precipitate maintains its composition up to 185° it can be weighed as such after drying at 105–15°.

Procedure

Dilute the soln. containing titanium (10–25 mg) to 150 ml, add 10 N hydrochloric acid (15–20 ml), heat to 70–80° and add with stirring an ethanolic reagent soln. (2–3 times the stoichiometric quantity) when a lemon-yellow precipitate appears. Heat to boiling, allow to stand at room temp. for 2 hr, and filter through a No. 4 sintered glass crucible. Wash several times with cold aqueous 0·05% reagent soln., dry at 105–15° and weigh.

Weight of titanium complex × 0·07931 = weight of titanium.

Effect of Other Ions

During the precipitation of titanium (10–15 mg) by the above procedure, the cations which do not interfere when present in the mg amounts given in parentheses are: thallium(I) (250), copper (150), beryllium (200), magnesium (200), zinc (200), manganese (200), aluminium (200), cadmium, mercury, iron(II and III), cobalt, nickel, chromium, arsenic(III), antimony, bismuth, cerium(IV), zirconium, thorium, vanadium(IV) and uranium(VI) (all 100). Tungsten(VI), molybdenum(VI) and tantalum(V) interfere. The interference of niobium can be masked by oxalate. Of the anions, 500 mg each of oxalate, tartrate, citrate, borate, acetate and EDTA and 200 mg of phosphate do not interfere, but 100 mg of thiocyanate and 5 mg of fluoride do interfere. Boric acid, however removes the effect of fluoride.

6. Applications of Thiobenzoylphenylhydroxylamine

DETERMINATION OF COPPER
(Cassidy and Ryan[38])

Thiobenzoylphenylhydroxylamine (TBPHA), which is insoluble in water but soluble in strongly acidic or weakly basic solutions, precipitates copper completely in the presence of EDTA at pH 2·0–2·5 as a red-brown complex of composition $Cu(C_{13}H_{10}OSN)_2$. To eliminate contamination by the reagent, the solutions must be 35–75% by volume in ethanol. Precipitation made from solutions which are 75% in ethanol requires cooling of the suspension to room temperature before filtration. As iron, aluminium, nickel, cobalt, zinc, antimony, bismuth, lead and tin do not interfere, copper (10–30 mg) can be determined by the procedure given below in the presence of these commonly associated ions and so also in copper ores, silver alloy, Monel metal and brass.

Procedure

Adjust the pH of the soln. to be analysed to 2·0–2·5, dilute to 100 ml and add a 0·25 M EDTA soln. in sufficient quantity to complex all the interfering ions. Avoid adding a large excess of EDTA to prevent the possibility of precipitation of the free acid of EDTA on standing. Make the soln. 45% (v/v) in ethanol. Heat on a water bath to 70° and precipitate copper by adding dropwise a 1% TBPHA soln. in ethanol until the supernatant becomes yellow, due to the presence of excess reagent. Filter through a porcelain gooch crucible, wash with hot 20% aqueous ethanol, dry at 100° and weigh.

Weight of copper complex × 0·12217 = weight of copper.

Determination of Copper in Ores and Alloys

Dissolve a suitable weight of sample in conc. nitric acid, fume with conc. sulphuric acid, dilute with water to 50 ml, and adjust to pH 1·5 with 4 N sodium hydroxide soln. Filter and wash any precipitate present. From the combined filtrate and washings precipitate copper by the above procedure. For silver alloys, precipitate silver as its chloride before proceeding.

DETERMINATION OF IRON IN STEEL AND IRON ORE AND SEPARATION FROM COPPER
(Abraham, Abraham and Ryan[39])

With a 5:1 molar excess of reagent to metal, iron(III) precipitates quantitatively from solutions 1 F in hydrochloric acid or 3 F in sulphuric acid after digestion for 1 hr at 50°. The precipitate adheres to the walls of the beaker at higher temperatures. At room temperature, quantitative precipitation is obtained on standing for 5 hr. A shorter digestion period gives low results as the precipitate is then colloidal and passes through the filter paper. The precipitate is sufficiently insoluble in water, but is highly soluble in ethanol. It must be ignited to the oxide as the weighing form, because of contamination with reagent.

The pure iron(III) complex, $Fe(C_{13}H_{10}NSO)_3$, which melts at 156° and decomposes above 200°, is prepared by adding an ammoniacal reagent solution to an iron(III) solution and by washing the precipitate with ammonia to remove the excess reagent.

Determination of Iron

To the soln. (containing 5–50 mg of iron) add enough hydrochloric acid so that after the addition of the reagent the acidity is 1 M in a volume of 100–200 ml. Precipitate iron by adding TBPHA (1 % in 1 F ammonia soln.) until a 5:1 molar excess of the reagent has been added or until the supernatant appears to be cloudy white. Digest on a water bath at 50° for 1 hr, cool to room temp., filter through a Black Ribbon S & S 589 filter paper, wash with a saturated aqueous soln. of the reagent (approx. 0·01 %), using up to 50 ml of wash soln., ignite to the oxide and weigh.

Weight of iron oxide (Fe_2O_3) × 0·6994 = weight of iron.

Under the condition of the experiment, iron (15–21 mg) can be determined in the presence of 20 mg each of manganese, mercury(II), aluminium, chromium, uranium(VI) and thorium and 30 mg each of zinc, lead, nickel, cobalt and zirconium (if masked by fluoride). Phosphate, fluoride, tartrate and EDTA do not interfere.

Determination of Iron in a Standard Steel or Iron Ore

Dissolve a suitable weighed amount of a standard steel or of an iron ore in hydrochloric acid with the addition of a few drops of conc. nitric acid to oxidize iron to the tervalent state. Filter, then precipitate iron with TBPHA, after dilution and adjustment of the acidity to 1 F.

Separation and Determination of Copper and Iron

Add to the soln. to be analysed enough EDTA to mask iron(III). Adjust to pH 2 and precipitate copper from 45% ethanolic soln. with TBPHA as described above[38] and weigh as $Cu(C_{13}H_{10}NSO)_2$ Evaporate the filtrate to a minimum vol. to drive off the ethanol, dilute and precipitate iron with TBPHA in ammonia from the soln. 1 F in hydrochloric acid and determine as the oxide as described above.

References

1. SHOME, S. C., *Analyst* **75,** 27 (1950).
2. RYAN, D. E. and LUTWICK, G. D., *Can. J. Chem.* **31,** 9 (1953).
3. BLAKELEY, J. H. and RYAN, D. E., *Analyst* **89,** 721 (1964).
4. LYLE, S. J. and SHENDRIKAR, A. D., *Anal. Chim. Acta* **36,** 286 (1966).
5. ALIMARIN, I. P. and YUN-HSIANG, CHIEH, *Talanta* **8,** 317 (1961). Translated from *Zavod. Lab.* **12,** 1435 (1959).
6. SINHA, S. K. and SHOME, S. C., *Anal. Chim. Acta* **21,** 459 (1959).
7. SINHA, S. K. and SHOME, S. C., *Anal. Chim. Acta* **21,** 415 (1959).
8. ALIMARIN, I. P. and YUN-HSIANG, CHIEH, *Vestnik Moscow Univ. Khim.* **15**(2), 53 (1960).
9. DAS, B. and SHOME, S. C., *Anal. Chim. Acta* **33,** 462 (1965).
10. ALIMARIN, I. P. and YUN-HSIANG, CHIEH, *Zhur. Anal. Khim.* **14,** 574 (1959); *Talanta* **9,** 9 (1962).
11. RYAN, D. E., *Can. J. Chem.* **38,** 2488 (1960).
12. SINHA, S. K. and SHOMA, S. C., *Anal. Chim. Acta* **24,** 33 (1961).
13. DAS, J. and SHOME, S. C., *Anal. Chim. Acta* **24,** 37 (1961).
14. DAS, J. and SHOME, S. C., *Anal. Chim. Acta* **27,** 58 (1962).
15. DAS, H. R. and SHOME, S. C., *Anal. Chim. Acta* **27,** 545 (1962).
16. ALIMARIN, I. P. and HAMID, S. A., *Zhur. Anal. Khim.* **18,** 1332 (1963).
17. KAIMAL, V. R. M. and SHOME, S. C., *Anal. Chim. Acta* **31,** 268 (1964).
18. DAS, B. and SHOME, S. C., *Anal. Chim. Acta* **32,** 52 (1965).
19. DAS, B. and SHOME, S. C., *Anal. Chim. Acta* **35,** 345 (1966).
20. DAS, B. and SHOME, S. C., *Anal. Chim. Acta* **40,** 338 (1968).
21. KAIMAL, V. R. M. and SHOME, S. C., *Anal. Chim. Acta* **27,** 298 (1962).
22. KAIMAL, V. R. M. and SHOME, S. C., *Anal. Chim. Acta* **29,** 286 (1963).
23. MOSHIER, R. W. and SCHWARBERG, J. E., *Anal. Chem.* **29,** 947 (1957).

24. MAJUMDAR, A. K. and MUKHERJEE, A. K., *Naturwiss.* **44,** 491 (1957).
25. MAJUMDAR, A. K. and MUKHERJEE, A. K., *Anal. Chim. Acta* **19,** 23 (1958).
26. MAJUMDAR, A. K. and MUKHERJEE, A. K., *Anal. Chim. Acta* **21,** 245 (1959).
27. MUKHERJEE, A. K., Ph.D. Thesis, Jadavpur University, 1959.
28. MAJUMDAR, A. K. and MUKHERJEE, A. K., *Z. Anal. Chem.* **189,** 339 (1962).
29. LANGMYHR, F. J. and HONGSLO, T., *Anal. Chim. Acta* **22,** 301 (1960).
30. MAJUMDAR, A. K. and PAL, B. K., *Anal. Chim. Acta* **24,** 497 (1961).
31. PAL, B. K., Ph.D. Thesis, Jadavpur University, 1964.
32. ALIMARIN, I. P., SOKOLOVA, I. V. and SMOLINA, E. V., *Vestnik Moscow Univ. Khim.* **23**(1), 67 (1968).
33. CARDWELL, T. J. and MAGEE, R. J., *Microchem. J.* **13,** 467 (1968).
34. MAJUMDAR, A. K. and MUKHERJEE, A. K., *Anal. Chim. Acta* **22,** 514 (1960).
35. MAJUMDAR, A. K. and PAL, B. K., *J. Indian Chem. Soc.* **42,** 43 (1965).
36. GHOSH, N. N. and BHATTACHARYYA, A., *J. Indian Chem. Soc.* **44,** 972 (1967).
37. SAVARIAR, C. P. and JOSEPH, J., *Anal. Chim. Acta* **47,** 347 (1969).
38. CASSIDY, R. M. and RYAN, D. E., *Anal. Chim. Acta* **41,** 319 (1968).
39. Abraham, I. D., ABRAHAM, J. and RYAN, D. E., *Anal. Chim. Acta* **48,** 93 (1969).

CHAPTER 5

SPECTROPHOTOMETRIC DETERMINATION OF THE ELEMENTS WITH N-BENZOYLPHENYLHYDROXYLAMINE AND ITS ANALOGUES

As a spectrophotometric reagent, N-benzoylphenylhydroxylamine (BPHA) has been much more extensively investigated than its analogues. The species that have been studied are vanadium(V), iron(III), mercury(II), cerium(IV), zirconium, titanium(IV) and niobium, the latter two in conjunction with thiocyanate. Vanadium(V) forms two types of complexes, according to the conditions, one with maximal absorption at 440 nm and the other at 530 nm.

The reaction sensitivities of vanadium(V) with BPHA and its analogues at high acidities are N-cinnamoylphenylhydroxylamine (540 nm; having a conjugated system) > N-2-thiophenecarbonyl-p-tolylhydroxylamine (530 nm) > N-furoylphenylhydroxylamine (530 nm) > N-2-thiophenecarbonylphenylhydroxylamine (530 nm) > N-benzoyl-o-tolylhydroxylamine (510 nm) > N-benzoyl-p-chlorophenylhydroxylamine (530 nm) > BPHA (530 nm) > N-phenylacetylphenylhydroxylamine (510 nm). Of these, only N-benzoyl-o-tolylhydroxylamine behaves as a specific spectrophotometric reagent for vanadium. Most of these compounds on examination are found to form two types of vanadium complexes, like BPHA.

With BPHA, the iron(III) compound formed at a higher pH shows a broad maximum in the region 460–80 nm whereas that with titanium(IV) in the absence of thiocyanate is at 345 nm. Phosphate and fluoride prevent the reaction of iron(III) and titanium(IV), respectively, with the hydroxylamine derivative.

The mercury(II), zirconium and niobium complexes, after extraction into chloroform, are measured at 340–60 nm. The sensitivity for zirconium is higher than that for niobium at 350 nm. The cerium(IV)–BPHA complex in chloroform, however, shows a maximum at 460 nm.

A more intensely coloured, chloroform extractable 1:2:1 complex is formed by niobium on reaction with BPHA in presence of thiocyanate in a strongly acidic solution. The molar absorptivity is 46,500 at 360 nm. The colour developed on bringing thiocyanate into contact with the toluene extract of the niobium–BPHA complex, produced in presence of tin(II) chloride, however, has a lower molar absorptivity, 32,000 at 365 nm.

Titanium(IV), like vanadium(V), forms different types of complexes according to the acidity of the solution. For instance, above pH 1 it gives a 1:2 product with BPHA, the colour of which is less intense than that given at higher acidities, and is more susceptible to pH changes. If the hydrochloric acid concentration is increased to 2 N, a more intensely coloured, 1:4 species is formed with the absorption peak at 355 nm. In strongly acidic solution, the ratio of titanium to BPHA is 1:2, probably because of the formation of a ternary complex, which on isolation appears to be $Ti(C_{13}H_{10}O_2N)_2Cl_2$. The chloroform extract—of the complex has an absorption peak at 380 nm. An excess of chloride also keeps uranium(VI) in an anionic form unreactive to BPHA.

If thiocyanate is added before the extraction of the titanium–BPHA complex from hydrochloric or sulphuric acid into chloroform, the colour intensity increases considerably. This may be due to the replacement of the chloride in the ternary complex by thiocyanate. But in solution the composition of the complex is 1:2:1 (Ti:BPHA:SCN^-) and not 1:2:2. However, the composition in solution may be different from that in isolated complex. A similar effect of thiocyanate in intensifying the colour of the reaction products of titanium and niobium with N-cinnamoylphenylhydroxylamine and of titanium with N-acetylsalicylphenylhydroxylamine and N-benzoyl-p-tolylhydroxylamine has also been noticed.

N-Benzoylmethylhydroxylamine, however, in aqueous solution at pH 1 forms a yellow complex with titanium with an absorption peak at 410 nm. This reagent has an added advantage in that it is soluble in water as its sodium salt.

N-Furoylphenylhydroxylamine finds favour as a spectrophotometric reagent for vanadium(V) and titanium(IV). Tin(II) chloride is used to reduce vanadium(V) and iron(III) to lower valence states before the determination of titanium.

N-Acetylphenylhydroxylamine gives on reaction with iron under different pH conditions three differently coloured products. That formed at pH 1·8–3·5 is the best for colour measurements. It has a molar absorptivity of 3071.

The complex of iron(III) with N-cinnamoylphenylhydroxylamine shows two maxima, one at 360 nm and the other at 480 nm, the molar absorptivity at the latter wavelength being 3400.

The only reagent of this type that has been used for the determination of uranium(VI) is N-cinnamoylphenylhydroxylamine; it forms at high pH values a complex that is extractable into organic solvents. The molar absorptivity is 16,000 at 355 nm.

All spectrophotometric analyses involve the use of the Lambert–Beer law and the determination of sensitivities as described by Sandell[1] is useful. The optimal range according to Ringbom[2] and the per cent analysis error[3] according to Ayres should also be evaluated.

The composition of the complexes in solution as determined in a few papers have usually been made by Job's method of continuous variation[4] as well as by mole ratio[5] and slope ratio[6] methods; their dissociation constants have been determined by the methods of Job[4,7] and Harvey and Manning.[6]

1. N-Benzoylphenylhydroxylamine as a Spectrophotometric Reagent

DETERMINATION OF VANADIUM
(Shome[8])

The mahogany-red compound formed by vanadium(V) with BPHA is soluble in ethanol, acetic acid and benzene. The coloured product formed in the presence of ethanol at pH 2·5 shows maximal absorption at 480 nm. The readings, however, are taken at 510 mμ. The system obeys the Lambert–Beer law.

The absorbance of the solution, with a vanadium concentration of 10 mg per litre, changes with pH. Between pH 1·9 and 2·8 it remains constant and below pH 1·9 the colour fades gradually. The optimal pH for maximal colour development is 1·9–2·8.

For a 50-ml solution, with a vanadium concentration of 15 mg per litre, 10 ml of an 0·2% BPHA solution in ethanol are sufficient for full colour development. The addition of more reagent has no effect. The colour is stable at pH 2·6 for about 5 hr. The smallest quantity of vanadium that can be detected in 50-ml Nessler tubes is 0·33 mg per litre of solution.

Procedure

In a 50-ml volumetric flask, pipette 1 to 15-ml aliquots of a vanadate soln. (containing 50 μg of vanadium per ml), add dilute sulphuric acid to adjust the pH to 1·9–2·8 followed by an ethanolic 0·2% BPHA soln. (10 ml) and ethanol (15 ml). Make up to volume with distilled water, allow to stand for 10 min and measure the absorbance at 510 nm in a 1-cm cuvette.

Effect of Other Ions

In a 50-ml solution adjusted to pH of 2·4–2·6 by dilute sulphuric acid, 0·5 mg of vanadium can be determined in presence of the following ions (in mg per litre), with < 2% change in absorption: manganese (360), titanium(IV) (50), tartrate (500), oxalate (25), arsenate (500), phosphate (500), molybdate (120), tungstate (10) and chromate (15).

DETERMINATION OF VANADIUM IN TITANIUM TETRACHLORIDE
(Zharovskii and Pilipenko[9])

Vanadium(V) combines with BPHA over a wide range of acidity, and both vanadium(V) and titanium(IV) complexes with BPHA are extractable into chloroform practically within the same acid region. The vanadium extract is reddish brown with maximal absorption at 445 nm, whereas that of titanium is yellow with maximal absorption at 345 nm. As the absorbance of the titanium extract is small at 445 nm, vanadium can be determined in presence of a considerable amount of titanium at this wavelength. Moreover, it has been observed that fluoride prevents the reaction of titanium with BPHA, and that 12 M phosphoric acid containing fluoride also masks iron, which otherwise gives a red complex with BPHA, extractable into chloroform, and having an absorption spectrum similar to the vanadium complex. Phosphoric acid also decreases the pH of the aqueous phase and creates a more favourable condition for the reaction of vanadium(V) with BPHA.

Vanadium(V), in the presence of ethanol, at pH 3, combines with BPHA in a ratio of 1:2, corresponding with the formula $V_2O_3(C_{13}H_{10}O_2N)_4$. The molar absorptivity of the chloroform extract of the vanadium complex is 3600 at 440 nm. The corresponding values for the iron(III) and titanium(IV) complexes are 4450 at 440 nm and 5300 at 345 nm, respectively.

Procedure

Dissolve a weighed amount (1 g) of the material in (1 + 1) hydrochloric acid (10 ml). Transfer the soln. to a 50-ml flask and dilute with water to the mark. Transfer 5 ml of the soln. to a separatory funnel, add to it in succession sodium fluoride (0·3g), 12 M phosphoric acid (0·25 ml), a 5% BPHA soln. in ethanol (0·4 ml) and chloroform (5 ml). Shake for 5 to 10 sec. Separate the layers and measure the optical density of the chloroform layer in a 0·5-cm cuvette at 450 nm or with the help of a blue filter. Compare with a calibration graph prepared with known quantities of vanadium. The method allows 0·01–0·30% of vanadium(V) to be determined.

DETERMINATION OF VANADIUM IN HIGH SPEED STEEL, CHROME-VANADIUM STEEL, CHROMITE AND CHROME-MAGNESITE REFRACTORY
(Ryan[10])

The vanadium complex precipitated from sulphuric acid on the addition of an ethanolic BPHA solution is brown. On dissolution in chloroform an orange-yellow solution is formed having an absorption maximum at 450 nm, with a molar absorptivity of 3700. From hydrochloric acid on the other hand, the chloroform extract is purple with maximal absorption at 530 nm and a molar extinction coefficient of 4490. With increasing hydrochloric acid concentration, the colour fades. Stable colour extracts are obtained from solutions which are 5–9 M in hydrochloric acid. The colour system so produced obeys the Lambert–Beer law at 530 nm. When exposed to day light for about 4 hr, the optical density of the system decreases.

Job's method of continued variation shows the complex formed in solution to contain vanadium and reagent in a ratio of 1:2.

Vanadate Solution

Dissolve sodium vanadate in distilled water containing 1% conc. sulphuric acid. From this standard soln., prepare weaker solns. of suitable concentration by dilution.

Procedure

To the vanadate soln. (containing 0·03–0·40 mg of vanadium), add hydrochloric acid to make the soln. 5–9 M in acid (Note). Transfer the soln., or an aliquot of it with vanadium within the above limits to a separatory funnel, and add for every 0·1 mg of vanadium, 0·5% BPHA soln. in chloroform (2 ml) followed by chloroform (10 ml). Shake well and transfer the chloroform layer to a 50-ml flask. Add another 10 ml of chloroform to the main soln., shake and transfer the chlororm layer to the flask. Dilute the combined extracts in the flask with chloroform to the mark and measure the optical density in a 1-cm cuvette at 530 nm.

Effect of Other Ions

Vanadium (0·100–0·245 mg) can be determined in the presence of iron, chromium and aluminium (all 100 mg). Permanganate, dichromate, nitrite and nitric acid over 5%, must be absent. Molybdenum (up to 50 mg) present as ammonium molybdate, does not allow more than 90% recovery of vanadium even after four extractions. Smaller quantities of molybdenum, zinc, cobalt, nickel, copper, manganese, calcium, magnesium, titanium, zirconium, arsenic, tin, tungsten, silicon, phosphorus, carbon and sulphur as present in steels, chromite or chrome–magnesite refractories do not interfere.

Determination of Vanadium in High-speed and Chrome Vanadium Steels

Dissolve a weighed sample of steel (containing 1 mg of vanadium) in (1 + 4) sulphuric acid (75 ml) and add nitric acid (a few drops) to oxidize iron and tungsten. Filter through a sintered glass crucible and wash with hot water. Collect the filtrate and washings in a 100-ml volumetric flask and dilute to the mark. Transfer a 25-ml aliquot to a separatory funnel, add to it a few drops of a 0·1 N potassium permanganate soln. till a faint pink colour persists and then adjust the acidity to the proper value with hydrochloric acid. Add the chloroformic 0.5% BPHA soln. extract, dilute with chloroform and measure the absorbance at 530 nm in a 1-cm cuvette as described above.

Determination of Vanadium in Chromite and Chrome–Magnesite Refractories

Fuse the sample (0.5g, containing 0.03–0.11% of vanadium) in a platinum crucible with a mixture of sodium carbonate and sodium tetraborate. Dissolve by warming the melt in (1 + 4) sulphuric acid (50 ml). Transfer to a volumetric flask and dilute to 100 ml. Pipette 25 ml of the soln. into a separatory funnel. To reduce chromium(VI) to chromium(III), add iron(II) sulphate soln. till the soln. is green. Add a 0.1 N potassium permanganate soln. till the soln. is just pink, adjust the acidity with hydrochloric acid, extract with the chloroformic BPHA soln. and determine vanadium as described above.

Alternatively, fuse the sample with sodium peroxide in a nickel crucible. Extract the melt with water, boil for 10 min, and cool. Add ammonium carbonate (5 g), filter, wash the precipitate with hot water, and neutralize the filtrate with (1 + 1) sulphuric acid. Transfer the soln. to a 100-ml volumetric flask, dilute to the mark and then proceed to determine vanadium from an aliquot portion as stated above.

Note. The hydrochloric acid should preferably be between 2.8 and 7.0 M so that the Lambert–Beer law is obeyed by 0.7–12 μg of vanadium per ml.

EXTRACTIVE SEPARATION OF VANADIUM FROM VARIOUS IONS AND ITS DETERMINATION
(Priyadarshini and Tandon[11])

The colour system formed on extraction of the vanadium complex into chloroform shows maximal colour development when the hydrochloric acid concentration is between 2.8 and 4.3 M. The sensitivity of the colour reaction according to Sandell is then 0.011 μg of vanadium per cm^2 at 510 nm, the wavelength of maximal absorption. The molar absorptivity is 4650. The colour system is stable for several days. Even a temperature variation between 20° and 40° has no effect. BPHA must be present in a molar ratio of 10 to 1 (BPHA:vanadium) to ensure full colour development. Under the stated conditions, the system obeys the Lambert–Beer law at 510 nm from 0.7 to 12 μg of vanadium per ml, the practical range being 2 to 8 ppm. But as the hydrochloric acid concentration of the solution is decreased, the wavelength of maximal absorption shifts to shorter wavelengths until at a concentration of 0·01 M hydrochloric acid, the extract is mahogany red with maximal absorption at 440 nm (molar absorptivity 3200). Such a shift of the absorption peak is also observed on the addition of alcohol to the chloroform used for extraction.

Reagents
BPHA solution. 0·1 % in alcohol-free chloroform.

Chloroform, alcohol-free. Wash the chloroform 5–6 times with half its volume of water, dry over fused calcium chloride, distil and store in an amber bottle.

Procedure
Add distilled water and 6 M hydrochloric acid to the vanadate soln. (containing 0·02–0·2 mg of vanadium) in a separatory funnel, so that the volume of the soln. is 25 ml and its acid strength is 2·8–4·3 M. Add BPHA soln. (10 ml), shake vigorously and allow the phases to separate for 2 min. Collect the chloroform layer in a 50-ml beaker containing anhydrous sodium sulphate (1·5 g). Add a further two 5-ml portions of chloroform to extract any violet complex remaining and collect the extracts in the beaker. Decant the violet chloroform soln. into a 25-ml volumetric flask. Wash out any adhering colour from the crystals of sodium sulphate with small portions of chloroform. Transfer these washings to the 25-ml flask and make up to volume with chloroform. Measure the optical density at 510 nm in a 1-cm cuvette against 0·1 % BPHA in chloroform.

Alternatively, for routine work, centrifuge the chloroform extracts to remove any tiny drops of water. Transfer to a 25-ml flask, make up to volume with chloroform and measure its absorbance as described above.

Effect of Other Ions
Vanadium(V) (91·5 μg) can be determined in the presence of aluminium, chromium(III), iron(III), manganese(II), copper(II), cobalt, nickel, zinc, thorium, citrate, tartrate, phthalate, phosphate, perchlorate, nitrate and sulphate even when the weight ratio of each of these ions to vanadium is 225 to 1. Since uranium(VI) consumes the reagent under the experimental conditions, a large excess of the reagent is required for the determination of vanadium in presence of uranium. Titanium, molybdenum(VI), zirconium, tungsten(VI), lead, mercury(I), silver, thallium(I), oxidizing agents, which oxidize the reagent, and reducing agents, which reduce vanadium(V) to lower valence states, must be absent.

DETERMINATION OF VANADIUM AND ITS SEPARATION FROM IRON
(de Pool and Cadavieco[12])

The coloured vanadium complex formed in 50 % ethanol, when the pH is adjusted with dilute sulphuric acid to 2·4–4·4, can be measured at 450 nm. The precision is 0·024 mg of vanadium per ml. In the presence of iron(III), however, the vanadium complex must be extracted from 3 M hydrochloric acid into benzene, before the absorbance measurement at 530 nm.

Procedure

Add to the vanadate soln. in a 25-ml calibrated flask a sufficient excess of an 0·2% BPHA soln. in 50% ethanol. Adjust the pH to 2·4–4·4 by the addition of dilute sulphuric acid, make up to volume with ethanol (to prevent the appearance of turbidity), wait for 10 min and measure its absorbance.

Determination in Presence of Iron(III)

To the test soln., add an excess of an 0·2% BPHA soln. in 50% ethanol, followed by conc. hydrochloric acid and distilled water to make the final (25 ml) soln. 3 M in hydrochloric acid. Extract the complex twice with 5-ml portions of benzene and transfer the benzene layers to a 25-ml flask. Make up to volume with benzene and measure the absorbance at 530 nm in a 1-cm cuvette.

EXTRACTIVE DETERMINATION OF VANADIUM IN STEEL AND PETROLEUM PRODUCTS
(Tomioka[13])

The colour system formed on the extraction of the vanadium complex from sulphuric acid is more stable, but has a smaller absorbance when compared with that extracted from hydrochloric acid. Of the solvents tested as extractants, although chloroform appears to be favourable, a considerable increase in sensitivity is observed with the use of a mixture of ethanol and chloroform in a ratio of 1:4. The sensitivity of reaction is then 0·013 μg of vanadium per cm^2. Though iron, titanium, molybdenum and tungsten interfere, the method, as developed for the determination of vanadium, is applicable to the analysis of petroleum products and to some steels when phosphoric acid is used before extraction to mask iron.

Procedure

To the sample soln. (containing 5–150 μg of vanadium(V)) which is 1·5 N in sulphuric acid, add phosphoric acid (2–5 ml) and dilute to 50 ml with

1·5 N sulphuric acid. Add an 0·2% BPHA soln. in the 1:4 ethanol–chloroform mixture (10 ml, sufficient for 150 μg of vanadium). Shake to extract the coloured complex into chloroform and measure its absorbance at 440 nm.

DETERMINATION OF TRACES OF VANADIUM IN ALUMINIUM
(Antonijevic[15])

Dissolve the sample (0·5 g) by heating in a mixture of (1 + 1) nitric acid (40 ml), perchloric acid (6 ml) and a few drops of hydrofluoric acid. Evaporate to dryness, dissolve in water and add drops of 0·1 N potassium permanganate soln. until the soln. is distinctly pink. Boil for 10 min, cool, and add conc. hydrochloric acid to make the soln. 4 N in the acid. From this soln. take an aliquot containing ≤ 0·0125 g of aluminium per ml, extract twice with 10-ml portions of a 0·1% BPHA soln. in chloroform and measure the absorbance of the extract at 530 nm. Prepare a calibration curve with the same concentration of aluminium. For 10–200 μg of vanadium, the error is ± 5%.

DETERMINATION OF SMALL AMOUNTS OF VANADIUM IN ROCKS
(Patrovsky[15])

Small amounts (0·001–0·015%) of vanadium in rocks have been determined by extraction into chloroform of the vanadium(V)–BPHA complex in the presence of sulphosalicylic acid and fluoride as masking agents.

Procedure

Weigh out the finely powdered sample (0·5–1·0 g) in a platinum crucible and decompose by heating with conc. hydrofluoric acid (5–10 ml) and (1 + 1) sulphuric acid (0·5 ml). Evaporate to fumes of sulphuric acid. Dissolve the residue in (1 + 1) hydrochloric acid (10 ml), filter, wash with water and evaporate the combined filtrate and washings to dryness. Add conc. hydrochloric acid (12–18 ml) and water (25 ml) to the dried residue and heat gently to dissolve. Oxidize vanadium to the pentavalent state by the addition of a few drops of a 5% potassium permanganate soln. until the soln. is dark violet and heat until colourless. Add sulphosalicylic acid (0·1 g) and ammonium fluoride (0·1–0·3 g) to the cold soln. and extract vanadium by shaking for 30 sec with an 0·1% chloroformic BPHA soln. (10 ml). Filter, and measure the absorbance of the extract at 530 nm.

DETERMINATION OF VANADIUM IN PLUTONIUM–VANADIUM ALLOYS
(Baughman and Waterbury[16])

BPHA has been used for the development of a rapid method for the determination of vanadium in plutonium–vanadium alloys. From 2·8–4·2 M hydrochloric acid, vanadium(V), in concentrations between 120 ppm and 5% in the alloys, is separated from 100 mg of plutonium(VI) by the extraction of the vanadium–BPHA complex into chloroform. The absorbance of the extract is measured at 530 nm (molar absorptivity, 4260).

Preparation of Sample

Weigh out accurately plutonium metal or alloy turnings (0·5–1·0 g) into a 40-ml centrifuge tube, add water (1 ml) and a few drops of 12 M hydrochloric acid to dissolve the sample. Centrifuge for 5 min. If there is no residue, transfer the soln. to a volumetric flask of such a volume so as to give a concentration of 4–120 μg of vanadium per ml on dilution to make up its volume. Dilute to the mark with 3 M hydrochloric acid. If there is any residue, decant the soln. into a flask of proper volume and put the residue in a platinum dish. Add a few drops of 70% perchloric acid and heat to evaporate hydrochloric acid. Add 15·7 M nitric acid (4 ml), 48% hydrofluoric acid (5 drops) and 70% perchloric acid (0·5 ml). Heat under an infra-red lamp over a hot plate to dense fumes. Cool, add a few drops of water and hydrochloric acid and warm to dissolve any solids. Transfer the soln. into the volumetric flask. Repeat the fuming operation if there is any residue. Finally, dilute to volume with 3 M hydrochloric acid.

Determination of Vanadium

From the soln, pipette an aliquot containing 12–225 μg of vanadium and ≯ 0·1 g of plutonium into a small beaker. Add to it 15·7 M nitric acid (0·5 ml) followed by 70% perchloric acid (1 ml). Heat the beaker covered with a watchglass on a hot plate at 225° under the infra-red lamp until it has almost completely evaporated with copious fumes. Remove from the hot plate, cool the beaker and add distilled water (1 ml) cautiously without spattering. Wait for 1 min, wash the watch glass with distilled water, add the washings to the beaker, cool and then wash the cold soln. with 12 M hydrochloric acid (3 ml) into an extraction vessel and dilute with water to 10 ml. Add an 0·5% chloroformic BPHA soln. (5 ml) and close the extraction vessel with a rubber stopper provided with one hole. Into the hole insert a 5-cm long glass tube having an internal diameter of 8 mm and connect this through a stop-cock to vacuum. Suck in air by adjusting the stop-cock through the extraction vessel to mix the layers for 5 min, adding chloroform through 1-mm capillary tube to keep the volume of organic layer constant. Remove the glass tube, insert the end of a 10-ml Luer syringe and withdraw the plunger of the syringe to the 10-ml level to draw the liquid entirely from the capillary tube. Allow the phases to separate for 5 min and push out the organic phase into a dry, 10-ml volumetric flask. Repeat the extraction procedure twice more by adding chloroformic 0·5% BPHA soln. (3 ml, 2 ml respectively) to the extraction vessel, allowing 3 min extraction time in each instance. Combine the organic phases in the same volumetric flask, dilute to 10 ml with chloroform and measure the absorbance of the extract at 530 nm in a 1-cm cuvette using acetone or chloroform as a reference soln. Measure the absorbances of the reagent blanks, prepared in the same way as above but without the metal ions and subtract the average absorbance from that of the sample to get the correct value.

Effect of Other Ions

Vanadium (30 μg) can be determined in presence of 100 mg of each of the following ions: fluoride, chloride, bromide, iodide, nitrate, alkali metals, alkaline earth metals, beryllium, zinc, cadmium, mercury(II), nickel, rhenium(II), aluminium, arsenic(III), bismuth, cerium(III), gallium, indium, scandium, yttrium, lanthanum, germanium, ruthenium(IV) and plutonium(IV).

Titanium(III), niobium(V), tantalum, tungsten(VI), tin(IV), platinum(II), manganese(II) and chromium(III) interfere even when present at the 1-mg level.

DETERMINATION OF VANADIUM IN MAGNETITE, ILMENITE, CHROMITE AND IGNEOUS ROCKS
(Iwasaki, Ozawa and Yoshida[17])

After the decomposition of the mineral, vanadium(V) is extracted from 6 N hydrochloric acid into a chloroformic BPHA solution and the absorbance of the extract is measured at 532 nm. Under the condition of the experiment, 0·05 mg of vanadium(V) can be determined in the presence of 50 mg of titanium, masked by fluoride, 0·2 mg of chromium(VI) and 3 mg of manganese.

Procedure

In a platinum crucible dissolve a weighed portion of the sample of ilmenite, magnetite or ordinary rock in the usual way by treatment with hydrofluoric and sulphuric acids. Evaporate to copious fumes of sulphuric acid and dissolve the residue in hot water. For chromite and chromite-rich igneous rocks, decompose a weighed sample by fusion with sodium peroxide, dissolve in dilute sulphuric acid, reduce the chromate by a soln. of iron(II) sulphate. Oxidize the excess of iron(II) with nitric acid, remove silica by treatment with hydrofluoric acid, fume with sulphuric acid and dissolve the residue in water.

To the soln. add drops of 0·1 N potassium permanganate soln. to oxidize vanadium to the pentavalent state followed by a 2% sodium fluoride soln. in sufficient quantity to mask the titanium. Adjust the acidity to 6 N in hydrochloric acid, add a chloroformic 0·067% BPHA soln. (15 ml) and shake for 30 sec. Filter the extract and measure its absorbance at 532 nm.

DETERMINATION OF VANADIUM IN ILMENITE AND RUTILE
(Pilkington and Wilson[18])

The vanadium(V)–BPHA complex that is extracted into chloroform from hydrochloric acid is violet and has a higher maximum molar absorptivity (at 530 nm) than that of the yellow-brown species extracted from sulphuric acid (at 450 nm).[10] But hydrochloric acid concentrations above 6 M cause some reduction, which, however, can be eliminated if the yellow vanadium–BPHA extract obtained from sulphate media is scrubbed with strong hydrochloric acid. This technique also converts the vanadium complex to the more sensitive violet chloride form.

Owing to the relatively high stability of the titanium–BPHA complex, the interference of titanium, when the latter is present in a ratio of 250:1 (Ti:V), cannot be masked by the addition of complexing agents to the original extraction mixture. But by scrubbing the organic phase with an 8% pentasodium triphosphate solution in 2 M sulphuric acid its effect can be controlled.

To oxidize vanadium in typical ilmenite solutions containing much iron and titanium, heating at 95° with an excess of potassium permanganate for 15 min is found to be desirable.

Procedure

Add 200-mesh ilmenite (1 g) to sodium hydrogen sulphate (15 g) fused in a quartz flask. For rutile, add 0·5 g to sodium hydrogen sulphate (10 g). Fuse until the sample is completely decomposed, cool and dissolve the melt in 2 M sulphuric acid (50 ml) by heating. Filter into a 200-ml volumetric flask. If necessary, treat the residue in the same way as above, and make up to volume with water.

To a 20-ml aliquot in a 50-ml beaker, add 10 M sulphuric acid (3 ml) to maintain the acidity at 2 M. Heat at 95° for 20 min to reduce the volume to 20 ml. Add an 0·1% potassium permanganate soln. (2 ml), heat at 95° for 15 min, add dropwise a 1% sodium azide soln. to remove the excess of permanganate and boil to remove the excess of azide. Cool, transfer to a 100-ml separatory funnel, add 0·1 ml of the permanganate soln. and wait for 5 min.

Add a chloroformic 0·1% BPHA soln. (20 ml) extract by shaking for 2 min, allow to settle and shake again for 1 min with 20 ml of 11 M hydrochloric acid. Transfer the organic phase to another separatory funnel (retain the aqueous phase). Add a mixture consisting of 10% sodium triphosphate soln. (20 ml) and of 10 M sulphuric acid (5 ml). Shake for 2 min, transfer the organic phase to another separatory funnel and shake again for 1 min with 20 ml of 6 M hydrochloric acid. Transfer the organic layer to a 25-ml flask. Shake the three retained aqueous phases, each for 1 min, with the BPHA soln. (5 ml) and mix this extract with that in the flask. Dilute to 25 ml with chloroform and measure its absorbance at 535 nm in a 1-cm cuvette (molar absorptivity 4750).

If the chromium content of the ilmenite sample is 0·5–5·0% Cr_2O_3, oxidize to chromium(VI) and remove by extraction with methyl isobutyl ketone from 1 M hydrochloric acid. Evaporate the residual soln. to fumes with sulphuric acid to make it ready for the above procedure.

DETERMINATION OF VANADIUM IN IRON ORES
(Hofer and Heidinger[19])

The method outlined is claimed to be suitable for the determination of 0·05–0·25% of vanadium(V) in iron ores. The vanadium in the sample is oxidized to the pentavalent state in boiling 1·8 N sulphuric acid by permanganate and the vanadium–BPHA complex subsequently formed in 3 N hydrochloric acid is extracted into chloroform.

Procedure

Weigh out the finely powdered sample (0·2 g) in a 250-ml beaker, add conc. hydrochloric acid (20 ml) and (1 + 1) sulphuric acid (5 ml) and heat on a sand bath to dissolve the sample. Add conc. nitric acid (2 ml) to oxidize the iron to the tervalent state and evaporate to fumes. To the residue add water (30 ml) followed by (1 + 1) sulphuric acid (5 ml) and boil until a clear soln. is obtained. Dilute to 50 ml in a graduated flask. Pipette out 20 ml of the soln. into a 250-ml beaker and add to it 0·1 N potassium permanganate soln. until the colour is pink. Boil for 3 min and, if the colour fades, add 1–2 drops more of the permanganate soln. to restore the colour. Cool, transfer the soln. into a 100-ml separating funnel, washing with a small quantity of water, and maintain the volume at 25 ml by the addition of water as needed. Add 20% sodium nitrite soln. (1 drop) to remove the excess permanganate followed by urea (1 g) to remove the excess of nitrite. Add (1 + 1) hydrochloric acid (25 ml) to maintain its strength at 3 N and then a chloroform soln. 0·1% in BPHA (20 ml). Shake to extract for 1 min, filter through "white band" filter paper and measure the absorbance of the extract in a 1-cm cuvette at 530 nm against a reagent blank.

DETERMINATION OF IRON
(Yun-hsiang[20])

Iron can be determined spectrophotometrically with BPHA in a 50% acetone–water mixture. In the pH range 4–7, the colour system obeys the Lambert–Beer law from 0·5 to 10·0 μg of iron per ml. The sensitivity of the colour reaction is close to that obtained for sulphosalicylic acid in its reaction with iron.

Procedure

To the iron(III) soln. (containing 12·5–250·0 μg of iron) add 0·25% BPHA soln. in acetone (3 ml) (Note 1). Adjust to pH 4–7 by adding dilute ammonia soln. Transfer to a 25-ml flask. Make up to volume with the required volumes of acetone and water to keep the final soln. 50% acetone. Measure the absorbance in a 1-cm cuvette at 470 nm (Note 2) against the BPHA soln. as a blank.

Notes

1. 2 ml of a 0·25% BPHA soln. is sufficient to produce the maximal colour intensity.
2. The colour system gives a broad absorption band with a maximum from 460 to 480 nm.

DETERMINATION OF IRON IN SILICATE MATERIALS
(Ishii and Einaga[21])

Iron(III) is extracted from hydrochloric acid adjusted to pH 3 by a monochloroacetic acid–sodium monochloroacetate buffer, by a benzene solution of BPHA and the absorbance is measured at 440 nm. The molar absorptivity of the complex is 5380 ± 50. The elements which are usually found in silicate materials do not interfere, except for titanium and vanadium which must be removed prior to the extraction of iron(III).

The hydrochloric acid solution of iron is prepared after the decomposition of a weighed sample of the silicate material in hydrofluoric acid in the presence of sulphuric acid according to the usual procedure.

EXTRACTIVE DETERMINATION OF TITANIUM
(Zharovskii, Shpak and Piskunova[22])

The titanium–BPHA complex is extracted effectively by chloroform from 2 to 12 N hydrochloric acid. The 1:4 complex formed in 2 N hydrochloric acid on extraction into chloroform shows maximal absorption at 355 nm with a molar absorptivity of 5200. To avoid the interfering effect of zirconium, the absorption measurements for titanium determination are made at 360 nm. The absorbance of the chloroform extract of the titanium complex does not change for at least 5 hr.

The 1:2 complex formed by titanium with BPHA at pH 1 has less absorbance and is more pH sensitive, so is not suitable for spectrophotometric determination. The complex isolated at pH 1–2 appears to be $TiO(C_{13}H_{10}O_2N)_2$.

Procedure

To 100 ml of test soln., which is 2 N in hydrochloric acid and contains 0·12–1·00 mg of titanium, add a 5% alcoholic BPHA soln. (2 ml) and chloroform (10 ml). Shake for 30 sec and transfer the extract, after filtration through a dry filter paper, to a 25-ml graduated flask. Twice repeat the extraction with the addition of alcoholic BPHA soln. (1 ml) and chloroform (5 ml). Collect the extracts in the flask and make up to volume with chloroform. Measure the absorbance in a 3-cm cuvette at 360 nm. Prepare a calibration graph by taking known amounts of titanium through the whole procedure.

Effect of Other Ions

Qualitative studies show that aluminium, tin, antimony, tantalum, tungsten, and alkali and alkaline earth metals do not interfere in the determination of titanium; hence the method can be used for the determination of titanium in aluminium and its alloys.

DETERMINATION OF TITANIUM IN PRESENCE OF NIOBIUM, TANTALUM, TUNGSTEN AND ZIRCONIUM
(Schwarberg and Moshier[23])

In a 10^{-2} M tartaric acid solution, 60% in ethanol, titanium(IV) gives a coloured complex with BPHA at pH 1.8 ± 0.1 adjusted with sulphuric acid. The alcoholic solution containing 7 mg of BPHA obeys the Lambert–Beer law for 3 to 160 µg of titanium per 10 ml at 340, 370 and 400 nm. The continuous variation method indicates titanium to have formed a 1:1 complex with BPHA.

Solutions

Tartaric acid: 5.33×10^{-2} M. Dissolve tartaric acid (3 g) in 3% sulphuric acid (100 ml). Dilute 6.66 ml of this soln. to 25 ml.

Titanium(IV): 6.26×10^{-3} M. Fuse titanium dioxide (50 mg) with 15–25 times its weight of potassium pyrosulphate. Dissolve the melt in a small quantity of water containing conc. sulphuric acid (3 ml) and tartaric acid (3 g). Dilute to 100 ml with boiled distilled water.

Procedure

Take 6.66 of titanium soln. and dilute to 25 ml in a graduated flask.

To the titanium(IV) soln. (2 ml) in a 10-ml flask, add 0.5% sulphuric acid (2 ml) followed by ethanolic 0.2% BPHA soln. (3.5 ml). Dilute to the mark with ethanol.

To prepare a reference solution, add to the tartaric acid soln. (2 ml) in a 10-ml flask the same amount of sulphuric acid, BPHA and ethanol as in the sample soln.

Shake the soln. and after 5 min measure the absorbance of the sample soln. against that reference soln. in 1-cm cuvettes. If the absorbance of the sample soln. exceeds 0.80 at 325 nm, 0.75 at 340 nm, 0.35 at 370 nm or 0.12 at 400 nm, take a smaller volume of the titanium soln. and add to it the tartaric acid soln. to give a volume of 2 ml, and then proceed as directed above.

Effect of Other Ions

Under the specified conditions, titanium (10–80 µg) can be determined in presence of niobium (160 µg) at 370 nm, tantalum (80 µg) at 370 nm, zirconium (80 µg) at 400 nm and tungsten (120 µg) at 370 nm with an accuracy of ± 0.8 µg. Iron(III), vanadium(V) and molybdenum(VI) must be absent.

DETERMINATION OF TITANIUM IN FERROUS AND NON-FERROUS ALLOYS AND IN THE PRESENCE OF TIN, ZIRCONIUM, TUNGSTEN, VANADIUM AND MOLYBDENUM
(Tanaka and Takagi[24])

The titanium(IV)–BPHA complex on extraction into chloroform from > 9·6 N hydrochloric acid has an absorption maximum at 371 nm, where the absorption of the chloroformic BPHA solution is very slight. The colour system is stable and obeys the Lambert–Beer law, both at 371 and 380 nm, up to 100 μg of titanium; the molar absorption coefficient is 6700 at 371 nm and 6600 at 380 nm. It is preferable to measure the absorbance of the coloured complex at 380 nm, because the absorbance of the reagent solution at this wavelength is lower. Mole ratio and continuous variation methods indicate that titanium combines with BPHA in concentrated hydrochloric acid to form a 1:2 complex.

Titanium solution. Fuse titanium dioxide (0·167 g) with potassium hydrogen sulphate and dissolve the melt in dilute hydrochloric acid (100 ml). Prepare weaker solns. by dilution of an aliquot with conc. hydrochloric acid.

Procedure

To an aliquot of the titanium soln. (containing ≯ 100 μg of titanium) in a separating funnel, add sufficient conc. hydrochloric acid so that the acidity of the final soln. (10–30 ml) is > 9·6 N. Add 0·1% BPHA in chloroform (10 ml), shake for 1–2 min, filter the extract through a filter-paper that has been well dried in a desiccator and measure the absorbance of the extract in a 1-cm cuvette at 380 nm against a reagent blank similarly treated.

Effect of Other Ions

Under these conditions, the presence of iron(II), tin(II and IV), zirconium, tungsten, vanadium(IV) in amounts up to several mg, and molybdenum(VI), if reduced by tin(II) chloride, is not harmful. Concentrated phosphoric, perchloric and sulphuric acids, each 1 ml per 10 ml of the aqueous layer, have practically no effect. Nitric or nitrous acids, which oxidize the reagent, must be absent.

Determination of Titanium in Ferrous Alloys

Dissolve the sample (1 g) in conc. hydrochloric acid or *aqua regia*. Add macerated filter paper, filter and wash the residue with hot water. Retain the filtrate in a 100-ml volumetric flask.

Ignite the filter paper holding any residue in a platinum crucible. Fuse with potassium hydrogen sulphate. Dissolve the melt in a little hot water, transfer the soln. to the 100-ml flask containing the filtrate and dilute to the mark with water.

Place a measured aliquot of this solution (containing < 100 μg of titanium) in a 100-ml beaker. Evaporate to a very small volume. If *aqua regia* has been used for the dissolution of the sample, evaporate to dryness twice with the addition of a few ml of conc. hydrochloric acid each time to ensure that nitric acid has been completely removed. Transfer the residue to a separating funnel with conc. hydrochloric acid (20–30 ml). Add solid tin(II) chlroide in small amounts to reduce iron(III) to iron(II). Extract titanium and measure its absorbance as suggested above.

If the sample is non-ferrous and is soluble in hydrochloric acid, weigh out only as much sample as contains < 100 μg of titanium and then follow the procedure described for the ferrous alloys.

MICRO-DETERMINATION OF TITANIUM AND VANADIUM IN URANIUM COMPOUNDS
(Vita, Mullins and Trivisonno[25])

Separate methods for the determination of titanium(IV) and vanadium(V), each as low as 0·2 μg per g of uranium, have been developed. The absorbances of the extracts of the BPHA complexes of titanium(IV) and vanadium(V) in chloroform, obtained from 8 and 4 M hydrochloric acid, respectively, are measured at 420 nm for titanium and 530 nm for vanadium. By this extraction procedure, titanium or vanadium is also separated from other interfering ions. The presence of an excess of chloride keeps uranium(VI) in an anionic form unreactive to BPHA; moreover, the presence of sodium chloride aids the extraction of vanadium complex. Both titanium(IV) and vanadium(V) combine with BPHA to form 1:2 complexes.

Determination of Titanium

Weigh out a sample of uranium (1–3 g) into a 250-ml beaker. Add conc. perchloric acid (20 ml). If the sample is uranium hexafluoride or tetrafluoride, add boric acid (0·5 g per g of the fluoride). Heat to dissolve. Take the entire soln. or an aliquot containing 5–25 μg of titanium and heat to fumes until only 2–3 ml remain. Cool to 70–100°, add water (3–5 ml) and 8 M hydrochloric acid (50 ml). Transfer to a 125-ml separating funnel, wash the beaker with 8 M hydrochloric acid (5 ml) and add the washings to the main soln. Add an 0·07% iron(II) ammonium sulphate soln. (2 ml) to reduce any vanadium(V) present to vanadium(IV) and shake for 10 sec. Add chloroformic 0·2% BPHA soln. (20 ml), shake the mixture for 5 min and allow the phases to separate for 1 min. Filter the organic phase through a plug of glass wool, collect in a dry 50-ml centrifuge tube and measure its absorbance in a 5-cm cell at 420 nm against a reagent blank.

Effect of Other Ions

Copper(II), iron(II or III) and nickel(II) do not interfere. Thorium, zirconium and tungsten(VI) interfere slightly if present in quantities of 1 mg or more. Niobium, molybdenum(VI) (> 0·1 mg), vanadium(V), chromium(VI) and fluoride interfere.

Determination of Vanadium

Decompose a weighed amount (1–5 g) of the uranium sample in a 400-ml beaker with conc. perchloric acid (25 ml) with the addition of the same amount of boric acid if required as stated above for the determination of titanium. After dissolution, take the entire soln. or an aliquot, containing 5–25 μg of vanadium(V), and heat to fumes. Add sodium chloride (1 g) and fume again until the solution has a volume of 1–3 ml, in order to remove any chromium as chromyl chloride. Cool to 70–100°, add water (5 ml) and a 3·5 M sodium chloride soln. (20 ml). Stir, and add an 0·1 N potassium permanganate soln. dropwise until the soln. is pink.

Transfer the soln. completely to a 125-ml separating funnel using 8 M hydrochloric acid (30 ml) as a wash liquid. Stopper the funnel and shake it for 10 sec. Add the 0·2% BPHA soln. (20 ml), shake for 3 min to extract the vanadium(V) complex and stand for 1 min to allow the phases to separate. Filter the organic phase through a plug of glass wool into a dry 50-ml centrifuge tube and measure its absorbance in a 5-cm cell at 530 nm against chloroform as a blank.

Effect of Other Ions

Copper(II), iron(III), nickel, thorium, zirconium and tungsten(VI) and small quantities of sulphate, nitrate, fluoride and perchlorate do not interfere. Molybdenum(VI) (> 1 mg), titanium(IV), unless complexed with fluoride, and chromium(VI) interfere.

DETERMINATION OF TITANIUM IN THE PRESENCE OF INTERFERING IONS OTHER THAN NIOBIUM, IN STEELS, A SILICON–ALUMINIUM ALLOY AND IN BURNT REFRACTORIES

(Afghan, Marryatt and Ryan[26])

On the extraction with chloroformic BPHA solution (0·1%) of titanium (75·7 μg) from solutions containing hydrochloric, perchloric or sulphuric acids (2–4 N), no noticeable colour in the chloroform extract is observed. The colour intensity increases as the acid concentration is increased to 6 N, at which acidity the absorbance maximum is at 370 nm. However, maximal absorbance is attained at 10–12 N acid at 380 nm. Addition of sodium chloride to perchloric or sulphuric acid samples before extraction gives spectra similar to those obtained from hydrochloric acid solutions.

Addition of thiocyanate before extraction increases the molar absorptivity from 7100 to 15,400 in 10 N sulphuric acid and from 6700 to 12,000 in 10 N hydrochloric acid. But in 6·5–8·0 N hydrochloric acid, addition of a large excess of thiocyanate before extraction produces a complex with a molar absorptivity of 16,700.

The increased absorbance achieved in the presence of thiocyanate may be useful for the determination of titanium in non-ferrous materials. However, because iron, tantalum and zirconium interfere seriously at ~ 350 nm, the wavelength of maximal absorption in the presence of thiocyanate, the procedure involving the extraction of titanium from 10 N hydrochloric acid solution in the presence of tin(II) chloride, but in the absence of thiocyanate, is preferred for general use.

In the absence of tin(II) chloride, for a solution 5×10^{-5} M in titanium, 100-fold molar excess of nickel, chromium, iron, zirconium, vanadium, niobium, tantalum and molybdenum interfere. But in the presence of tin(II) chloride, only niobium interferes seriously. Moreover, common anions such as acetate, citrate, oxalate, tartrate, fluoride, nitrate and phosphate do not interfere even when present in 100-fold excesses.

Titanium complexes isolated from 10 N hydrochloric acid in the absence and presence of thiocyanate, appear to have the compositions $Ti(C_{13}H_{10}O_2N)_2Cl_2$ and $Ti(C_{13}H_{10}O_2N)_2(SCN)_2$, respectively. The compounds, however, were not of very high purity, although it is clear that a ternary complex titanium, BPHA and an anion is formed.

By continuous variation and molar ratio methods in hydrochloric acid and by the slope ratio method in sulphuric acid, titanium is found to form a 1:2 complex with BPHA in agreement with the results for the isolated complexes.

The chloroform extract of the ternary complex from 10 N hydrochloric acid, in the presence of tin(II) chloride, obeys the Lambert–Beer law from 1 to 6 ppm of titanium at 380 nm, the lower limit of determination being 0·25 ppm.

Determination of Titanium

Dilute the soln. to a known volume in a graduated flask. Transfer an aliquot (1 ml containing 10–50 μg of titanium) to a separating funnel, add a 5 M tin(II) chloride soln. (2·5 ml) in conc. hydrochloric acid and dilute to 25 ml with 10 N hydrochloric acid. Extract titanium from the aqueous phase with two 10-ml portions of a 0·1% soln. of BPHA in chloroform. Dilute the extract to 25 ml with chloroform and measure its absorbance at 380 nm in a 1-cm cell against a reagent blank.

For the determination of titanium in steels, a silicon–aluminium alloy and in burnt refractories, take a suitable weight (0·1–1·0 g) of the sample, prepare a soln. in hydrochloric or sulphuric acid by an appropriate procedure and determine titanium as above.

THE SOLVENT EXTRACTION AND DETERMINATION OF TITANIUM AS THE TITANIUM–BPHA–THIOCYANATE COMPLEX
(Che-ming and Shu-chuan[27])

A BPHA solution in chloroform or iso-amyl alcohol, when used for the extraction of titanium(IV) from a strongly acidic solution, gives a yellow complex suitable for the determination of mg quantities of titanium. However, the addition of ammonium thiocyanate markedly increases both the absorbance of the titanium–BPHA complex and its extractability into chloroform.

The absorption of the ternary titanium–BPHA–thiocyanate complex extracted into chloroform from 10–16 N sulphuric acid or 7–9 N hydrochloric acid is measured at 420 nm. The system obeys the Lambert–Beer law up to 20 μg of titanium, when the presence of only 10 mg of BPHA is enough to give maximal colour intensity. The system is stable for about 2 hr. By Job's method of continuous variation, titanium, BPHA and thiocyanate ion are found to be present in the molar ratio of 1:2:1 in the extracted complex.

Procedure

Adjust the acid concentration of the soln. (containing < 20 μg of titanium) to 10–16 N in sulphuric acid or to 7–9 N in hydrochloric acid. Add to the soln. (10 ml) ammonium thiocyanate (0·2 g) followed by an 0·2% soln. of BPHA in chloroform (5 ml). Shake to extract the coloured complex into chloroform, and measure its absorbance at 420 nm.

Effect of Other Ions

Under the conditions of the determination, up to 1 mg of cobalt(II), iron(III) and lanthanum may be present, and up to 100 μg of antimony, bismuth, nickel, tin(IV) and zirconium and up to 100 mg of aluminium, and tartaric, oxalic and phosphoric acids can be tolerated. Copper(II) gives low results but can be masked by thiourea. Vanadate, molybdate and tungstate do not interfere when present in up to 2–3 times the amount of titanium. Fluoride must be absent.

DETERMINATION OF NIOBIUM AS THE NIOBIUM–BPHA–THIOCYANATE COMPLEX
(Che-ming and Shu-chuan[28])

The niobium–BPHA complex, when extracted from sulphuric or tartaric acid into chloroform, is such a pale yellow that it is useless for the determination of μg amounts of niobium. But when a chloroformic BPHA solution is used to extract niobium from 7–8 N hydrochloric acid containing ammonium thiocyanate, the colour developed is intensified so that μg amounts of niobium can be determined. The system obeys the Lambert–Beer law for 3–30 μg of niobium in 5 ml of chloroform. The molar absorptivity of the complex is 46,500 at 360 nm, 30,700 at 380 nm and 18,000 at 420 nm. The niobium:BPHA:thiocyanate ratio is found to be 1:2:1. Extraction from 8–9 N sulphuric acid containing thiocyanate gives a 20% reduction in colour intensity compared with extraction from hydrochloric acid.

Procedure

To 10 ml of niobium soln. (containing 3–30 μg of niobium), adjusted to be 7–8 N in hydrochloric acid, add ammonium thiocyanate (0·2 g), followed by chloroform (5 ml, containing 0·4–0·7 mg of BPHA). Shake for 5 min, remove the chloroform layer and measure its absorbance at 380–420 nm.

Effect of Other Ions

Up to 0·3 g of aluminium, mg amounts of lanthanum and thorium and 0·05–0·1 mg of molybdenum, tungsten, zirconium and vanadium do not interfere. Tartrate and smaller amounts of oxalate and fluoride have no influence. Tantalum, titanium and larger amounts of sulphate must be absent.

DETERMINATION OF NIOBIUM IN URANIUM FISSION PRODUCTS AND IN THE PRESENCE OF OTHER IONS

(Villarreal and Barker[29])

The method developed is somewhat selective. The procedure is based on the extraction into toluene of the niobium–BPHA complex, formed in 9–12 N hydrochloric acid in the presence of tin(II) chloride and then shaking the extract with ammonium thiocyanate in 4 N hydrochloric acid. The molar absorptivity of the species thus formed in toluene is 32,000 and its sensitivity is 0·0029 μg of niobium/cm^2 at 365 nm. It obeys the Lambert–Beer law from 0 to 50 μg of niobium in 10 ml of toluene. The niobium–BPHA complex in toluene may also be used for the determination of niobium, but its molar absorptivity is only 10,000 at 365 nm.

The niobium–BPHA–thiocyanate colour forms within 1 min when extracting from an optimum acid concentration of 3–4 N in hydrochloric acid. It is stable for 1–2 hr, if kept in the dark. Exposure to direct sunlight quickly causes the formation of a yellowish turbidity, but when exposed to indirect sunlight, the absorbances of the sample and of the blank increase at the same rate. Thus the net absorbance remains constant for 1–2 hours.

Ammonium thiocyanate solution. Dissolve ammonium thiocyanate (250 g) in de-ionized water and dilute to 1 litre. Add isobutyl methyl ketone (100 ml), extract and discard the ketone layer. Add then chloroform (100 ml), shake to extract the ketone and discard the chloroform layer. Filter the soln. through a quantitative filter paper.

Dissolution and preparation of samples. For the dissolution of uranium fission products, take the sample (1 g) in a 125-ml Erlenmeyer flask, add to it de-ionized water (15 ml) and a 4:1 mixture of hydrochloric–nitric acids (20 ml). Heat on a hot plate to dissolve. Separate any undissolved residue, primarily due to ruthenium, by centrifugation in a 40-ml cone. Add 6 M sodium hydroxide soln. (2 ml), sodium hypochlorite soln. (5 ml) and heat on a water bath for 10 min to dissolve the residue. Cool, acidify with 12 N hydrochloric acid, add this to the original soln. and dilute to 50 ml with de-ionized water.

Fume strongly with conc. sulphuric acid the sample containing precipitated or polymerized niobium to less than 1 ml. Also, add conc. sulphuric acid (2 ml) to the sample aliquot which is more than 3 ml and fume to less than 1 ml. Transfer the residual soln. with 12 N hydrochloric acid (20 ml) to a separating funnel, and after the addition of a 40% tin(II) chloride soln. in 12 N hydrochloric acid (1 ml) follow the procedure given below, starting with BPHA addition.

Determination of Niobium

Transfer an aliquot not fumed with sulphuric acid (containing 2–20 µg of niobium) into a 60-ml separating funnel. Mix with 12 N hydrochloric acid (20 ml) and 40% tin(II) chloride soln. in 12 N hydrochloric acid (1 ml). Add a 1% soln. of BPHA in ethanol or acetone (1 ml), mix thoroughly, allow to stand for 1 min, add toluene (10 ml) and shake for 1 min to extract the niobium–BPHA complex formed. Allow the phases to separate. Drain off and discard the aqueous phase.

To the organic phase, add 4 N hydrochloric acid (20 ml), the 25% ammonium thiocyanate soln. (5 ml) and shake for 1 min. Allow to stand until the organic layer is clear. Drain through cotton a portion of the organic phase into a 1-cm cuvette and measure its absorbance against a blank within 1 hr after the addition of ammonium thiocyanate.

Effect of Other Ions

Amounts of 10 mg each of lithium, sodium, potassium, magnesium, calcium, barium, tin, boron, iron, yttrium, lanthanides, earths, thorium, molybdenum, uranium and ruthenium, and 1 mg each of copper, cobalt, nickel, zinc, manganese bismuth, aluminium, chromium, indium, titanium, silicon, tungsten, rhenium, platinum, palladium, and rhodium do not interfere.

The interference of up to 1 mg of zirconium and vanadium may be avoided by the use of 85% phosphoric acid (1 ml). Larger quantities of zirconium precipitate as the phosphate. Fluoboric acid (0·1 ml) can mask up to 10 mg of zirconium. Thiourea and thioglycollic acid (0·5 ml) mask the effect of 1 mg of silver and 10 mg of bismuth, respectively.

Masking agents, if needed, must be added to the hydrochloric acid (9–12 N) soln. of niobium, before the addition of BPHA.

Citrate, oxalate, tartrate, thiourea, thioglycollic acid, phosphate, perchlorate, sulphate, nitrate, bromide and iodide (all 100 mg) and fluoride (10 mg) have no interfering effect.

SEPARATION AND DETERMINATION OF ZIRCONIUM-95 AND NIOBIUM-95

(Shigematsu, et al. [30])

Zirconium-95 and niobium-95 can be extracted, respectively, from 0·05 to 0·10 N and 12–15 N sulphuric acid with a 98% and 95% recovery, by an 0·1% solution of BPHA in chloroform. Both systems obey the Lambert–Beer law up to 20 ppm. At 350 and 360 nm, the molar absorptivities for the zirconium chelate are 5200 and 2950, respectively. The corresponding values for the niobium chelate are 4250 and 3050.

Procedure

Adjust the acidity of a 10-ml sample soln. (containing 10–200 μg each of zirconium and niobium) to 12–15 N in sulphuric acid and add an 0·1% chloroformic BPHA solution (10 ml). Shake for 5 min to extract the niobium complex. Measure the absorbance of the chloroform layer within 30 to 50 min of extraction at 350 or 360 nm, against a reagent blank.

To determine zirconium (10–200 μg) follow the same procedure but maintain the acidity of the soln. between 0·05–0·10 N in sulphuric acid.

DETERMINATION OF MERCURY IN THE PRESENCE OF OTHER IONS
(Das and Shome[31])

The yellow product formed on extraction of mercury(II) with a chloroformic BPHA solution gives maximal absorption at 340 nm. With increasing amounts of mercury(II), the wavelength of maximal absorption increases slightly. Nevertheless, all absorption measurements can be made at 340 nm. The yellow species obeys the Lambert–Beer law at this wavelength for 5–52 μg of mercury per ml; the optimal range is 15–52 μg of mercury per ml. The absorbance is stable at 30° for about 4·5 hr, after which it gradually decreases. But even after 24 hr, the Lambert–Beer law is obeyed. At lower temperatures (5–10°) no appreciable change in absorbance is observed even after 1 week.

The optimal time for extraction of mercury (1·08 mg) with a chloroform solution of BPHA is 1 hr at pH 6·1–7·4. The extraction begins at pH 2·8 and the pH for 50% extraction is 4·8. The optimal BPHA concentration is 0·04 M. At BPHA concentrations below 0·01 M, extraction is incomplete even when the extraction time is 5 hr.

The sensitivity of the colour reaction is 0·075 μg/cm² and the molar absorptivity of the coloured species is 2693 ± 10.

Procedure

Take an aliquot of a mercury(II) nitrate soln. in 0·1 N nitric acid in a 100-ml separating funnel. Dilute to 30–40 ml with water. Add a 10% sodium acetate soln. to adjust the pH to 6·2. Extract with three successive 4-ml portions of an 0·8% BPHA soln. in distilled chloroform. Shake vigorously after each addition for about 20 min. Transfer the extracts to a

25-ml graduated flask. Make up to volume with chloroform and measure the absorbance at 340 nm in a 1-cm cuvette against a reagent blank.

Effect of Other Ions

Mercury (1·08 mg) can be determined in the presence of the following species (tolerance limits, in mg, are given in parentheses): silver (10), copper (2), lead (10), cadmium (1), zinc (1), manganese (1), bismuth (10), antimony (20), arsenic(V) (20), iron(III) (10), titanium(IV) (10), zirconium (10), thorium (20), tin(IV) (20), uranium(VI) (20), lanthanum (20), molybdenum(VI) (20), tungsten(VI) (20), vanadium(V) (10), fluoride (> 100), sulphate (> 100), oxalate (> 100), phosphate (20), borate (20), citrate (> 100) and tartrate (> 100). Chloride, cyanide and EDTA must be absent. To achieve these tolerance limits, silver, lead, bismuth, antimony, thorium, molybdenum(VI), uranium(VI) and lanthanum must be masked by citrate, tungstate and arsenate must be masked by tartrate, fluoride must be used to mask zirconium and titanium and tin(IV) must be masked by oxalate. Zinc (> 1 mg) must not be present as it forms an emulsion with the reagent. For the removal of iron(III), vanadium(V) and copper, extract the iron and vanadium at pH 2 and copper at pH 3·8 with a 0·01 M BPHA solution in chloroform.

DETERMINATION OF CERIUM
(Murugaiyan and Sankar Das[32])

The cerium(IV)–BPHA complex formed in a solution at pH 8–10 on the addition of the reagent at least 9 times the molar concentration of the metal ion is extractable into various organic solvents, but not into cyclohexane, heptane and hexane. The absorption spectra of the complex in non-polar solvents are similar. The chloroform extract, whose colour is stable for 20 days, shows an absorbance maximum at 460 nm, where it obeys the Lambert–Beer law over the range 4–40 ppm of cerium dioxide. Because ethanol reduces the percentage extraction, its use as a solvent for the reagent must be kept to a minimum. During the colour development, occasional stirring for 30 min converts any cerium(III) present to cerium(IV) by atmospheric oxidation.

Procedure

Dilute the cerium(IV) soln. (containing 0·5 mg of cerium dioxide) to 20 ml, add an 0·2% ethanolic reagent soln. and a 10% ammonium nitrate soln. (each 5 ml) and adjust the pH to 9. Stir occasionally for 30 min and transfer to a 100-ml separating funnel, rinsing the beaker twice with chloroform (5 ml). Shake for 3 min and collect the extract in a 25-ml graduated flask. Shake the aqueous phase twice with 3-ml portions of chloroform. Combine the chloroform phases in the flask, dilute to the mark with chloroform and measure its absorbance at 460 nm in a 1-cm cell against a reagent blank.

Effect of Other Ions

Alkaline earth metals, magnesium, zinc, cadmium, mercury(II), lead, lanthanides and thorium, at 20 times the amount of cerium, and chloride, sulphate, nitrate and acetate, even in 1000-fold excess, do not interfere. Tartrate, oxalate, citrate, fluoride, phosphate, carbonate, zirconium, cobalt, iron, titanium and uranium(VI) show positive interference while aluminium, copper, nickel, manganese and beryllium prevent complete extraction.

Many common elements may be extracted away as their BPHA complexes in the pH range 3–5 before the determination of cerium, e.g. in lanthanum oxide.

For the analysis of cerium glass and carbonates, the lanthanides may be precipitated as their fluorides, if need be, by the addition of lanthanum carrier, and converted to sulphates, the cerium(IV) content of which is then determined by the two-stage extraction procedure given below.

Determination of Cerium in Lanthanum Oxide

In a beaker, dissolve the sample (0·1 g, containing 0·1–0·6 mg of cerium dioxide) in 1 N hydrochloric or sulphuric acid (25 ml). Add a 25% ammonium chloride soln. (5 ml) and an 0·5% ethanolic BPHA soln. (2 ml). Dilute to 45 ml, adjust, using pH paper, to pH 4.5 ± 0.5, transfer to a 100-ml separating funnel and extract twice with 10-ml portions of chloroform. Discard the organic phase. Add to the aqueous phase BPHA (0·6 g) dissolved in ethanol (5 ml) and determine cerium by extraction into chloroform of its BPHA complex formed at pH 9.

2. N-Acetylphenylhydroxylamine as a Spectrophotometric Reagent

DETERMINATION OF IRON
(Gupta and Sogani[33])

Iron(III) can form three coloured products with N-acetylphenylhydroxylamine (APHA), depending on the pH. Below pH 1·2, the product is purple, with an absorption peak at 490 nm (Note 1). At pH 1·8–3·5, an orange species is formed that has a peak at 470 nm. The yellow product formed at pH 6·0–8·5 has an absorption peak at 425 nm. The colour development at pH 1·8–3·5 is instantaneous although for maximal colour development a 100-fold excess of the reagent is required. The colour so produced is stable at 34° for a reasonable period (Note 2) and obeys the Lambert–Beer law for 1 to 12 ppm of iron at 470 nm. The molar absorptivity is 3071; the sensitivity of the colour reaction as determined in Nessler tubes is $1:4\cdot5 \times 10^6$ parts with a detection limit of 0·2 µg of iron in 0·15 ml of solution.

Procedure

Pipette the iron(III) soln. (containing 25–300 µg of iron) into a 25-ml graduated flask. Add a 5% sodium potassium tartrate soln. (1 ml) and a suitable quantity of 10% hydrochloric acid to adjust the pH of the final soln. to 1·8–3·5. Add a 1% aqueous APHA soln. (10 ml) and make up to volume. Measure the absorbance of the soln. at 470 nm in a 1-cm cuvette against water as a blank.

Effect of Other Ions

For the determination of iron (2·8 ppm), silver, mercury(II), tin(IV), lead, cerium(IV), zirconium, uranium(VI), and alkali and alkaline earth metals are unlikely to interfere, because these ions do not give any colour reaction with the hydroxylamine derivative. Limiting concentrations (in ppm) of other ions are given in parenthesis: cobalt (120), nickel (320), copper (200), manganese(II) (200), zinc (200), cadmium (200), palladium (20), aluminium (160), chromium(III) (20), bismuth (40), antimony (40), titanium (50), platinum (IV) (12), phosphate (3000), tungstate (40) and cyanide (400). Gold(III) and vanadium(IV) must be absent (Note 3).

Notes

1. The purple colour formed at pH 1 is very unstable.
2. The orange colour is stable for about 1 hr.
3. Vanadium(V) also gives a bluish-violet colour and must be absent.

3. *N*-Benzoylmethylhydroxylamine as a Spectrophotometric Reagent

DETERMINATION OF TITANIUM
(Gupta and Sogani[34])

Titanium(IV) forms with *N*-benzoylmethylhydroxylamine (BMHA) in aqueous solution a deep yellow complex at pH 0·5–1·4 with an absorption peak at 410 nm. For full colour development, BMHA must be present in a large excess (about 340 times the amount of titanium). The colour develops instantaneously on the addition of BMHA solution. It shows no change in absorbance on varying the temperature between 20° and 30° and is stable for more than 18 hr. The coloured complex obeys the Lambert–Beer law over the range 0·5–6·5 ppm of titanium. The molar absorptivity is 4650. The sensitivity of the colour reaction as measured in Nessler tubes is $1:10^7$ and on a spot plate the detection limit is 0·2 μg of titanium in 0·15 ml of solution.

Reagents
BMHA. 2% aqueous soln. of *N*-benzoylmethylhydroxylamine (sodium salt).
Titanium(IV) solution. Fuse a weighed quantity of titanium dioxide (1 g) with potassium pyrosulphate (10 g), dissolve the melt in 6% (v/v) sulphuric acid (200 ml) and dilute to 500 ml. From this standard soln. prepare a soln. containing 25 ppm of titanium.

Procedure
To the titanium(IV) soln. (containing 15–150 μg of titanium) in a 25-ml flask, add a 10% sodium acetate soln. followed by a suitable quantity of hydrochloric acid and BMHA soln. (10 ml) to adjust the pH of the final soln. to between 0·5 and 1·4. Dilute with water to the mark and measure the absorbance 410 nm in a 1-cm cuvette, against a reagent blank.

Effect of Other Ions
Cyanide and tartrate have no effect on the colour reaction and mercury(II), tin(II), lead, cerium(IV), zirconium, uranium(VI) and the alkali and alkaline earth metals do not give a colour reaction. Tolerance limits of other ions (in ppm), as observed during the determination of 2 ppm of titanium, are given in parenthesis: platinum(IV) (6), chromium(III) (8), cobalt (30), manganese (40), aluminium (40), iron(II) (80), nickel, cadmium, tungstate, zinc and copper (each 120). Silver, gold(III) and vanadium(IV) must be absent. Tartrate can be used to mask tungstate.

4. *N*-2-Thiophenecarbonyl-*p*-tolylhydroxylamine and *N*-2-Thiophenecarbonylphenylhydroxylamine as Spectrophotometric Reagents
DETERMINATION OF VANADIUM
(Tandon and Bhattacharyya[35])

As spectrophotometric reagents, *N*-2-thiophenecarbonyl-*p*-tolyl-hydroxylamine (*p*-TTHA) and *N*-2-thiophenecarbonylphenylhydroxylamine (PTHA) behave similarly towards vanadium. An aqueous ethanolic solution of p-TTHA gives insoluble red and violet complexes with vanadium in solutions of pH 1·1–6·5 and 2–10 N acid, respectively. The violet complexes of both the reagents on extraction into chloroform from 2·8 to 5·0 N hydrochloric acid give maximal absorption at 530 nm, with molar absorptivities of 5750 $\pm$ 50 and 5450 $\pm$ 50 for the complexes with p-TTHA and PTHA, respectively. The sensitivities of the colour reactions, at 530 nm, are $8·8 \times 10^{-3}$ and $9·3 \times 10^{-3}$ μg of vanadium per cm^2, respectively, with p-TTHA and PTHA. Both reagents retain all the characteristics of BPHA as spectrophotometric reagents for vanadium, but are more sensitive. As both thiophene derivatives give practically the same results, those obtained with p-TTHA, which is slightly more sensitive, are described in detail.

The colour system formed on extraction into chloroform with p-TTHA as reagent obeys the Lambert–Beer law at 530 nm for 15–275 μg of vanadium per 25 ml with the optimal range of 44–155 μg of vanadium per 25 ml. The system, which is stable for several days when kept in a cool dark place, and is unaffected by slight variations of temperature, needs the presence of a 1:10 molar ratio of vanadium to reagent for full colour development. In actual experiments, 50 mg of the thiophene derivative is used per mg of vanadium.

Reagents

p-TTHA solution. 0·1% in chloroform (alcohol-free).

Alcohol-free chloroform. Wash the reagent grade chloroform 5–6 times with half its volume of water, dry over fused calcium chloride, distil and store in a dark place in an amber bottle.

Procedure

Treat the vanadate soln. with a few drops of bromine water to ensure the presence of vanadium(V), and boil off the excess bromine. Cool, transfer an aliquot (containing $\not>$ 0·15 mg of vanadium) to a separating funnel and add the requisite quantities of distilled water and 6 N hydrochloric acid to make the volume 25 ml and to adjust its acidity to between 2·8 and 5·0 N. Add the p-TTHA soln. (8–10 ml) and shake vigorously for 1 min. Allow the phases to separate for 2 min. Collect the chloroform layer in a small conical flask containing anhydrous sodium sulphate (1·5 g). To remove any vanadium complex left in the aqueous phase, add two more 5-ml portions of alcohol-free chloroform, shake and add the extracts to the previous chloroform extract. Decant the violet soln. from the conical flask into a 25-ml graduated flask and wash out with small portions of chloroform the adhering colour from the sodium sulphate crystals. Add the washings to the main soln. and dilute to the mark with alcohol-free chloroform. Measure the absorbance at 530 nm in a 1-cm cuvette against chloroform as the blank.

Effect of Other Ions

In the above procedure, for a vanadium concentration of 5·1 mg per litre, aluminium, chromium(III), iron(III), cerium(IV), thorium, uranium(VI), copper, cobalt, nickel, manganese, zinc, calcium, magnesium, nitrate, sulphate, perchlorate, phthalate, borate, citrate and tartrate do not interfere, even when each of these is present in a weight ratio of 250 to 1 of vanadium. Titanium, zirconium and molybdate must be absent. Silver, mercury(I), lead, tungstate and thallium(I), if present, give precipitates. The precipitates must be centrifuged off and washed before determining vanadium. If there is an appreciable amount of precipitate, the extraction procedure must be repeated several times.

5. N-Cinnamoylphenylhydroxylamine as a Spectrophotometric Reagent

DETERMINATION OF VANADIUM
(Priyadarshini and Tandon[36])

N-Cinnamoylphenylhydroxylamine (CPHA) forms a violet complex with vanadium in 2–10 N hydrochloric acid, which on extraction into chloroform shows a broad absorption band with its maximum at 530–50 nm. The colour development is maximal when the molar ratio of vanadium to CPHA is 1:10 and the solution is 2·7–7·5 N in hydrochloric acid. A large excess of reagent has no effect. In practice, about 70 mg of CPHA are used per mg of vanadium. The vanadium complex extracted into chloroform from 2·7–7·5 N hydrochloric acid solutions obeys the Lambert–Beer law for 0·5–10·0 ppm of vanadium at 540 nm, with optimal range from 1·5–5·0 ppm. The sensitivity is 0·008 μg of vanadium per cm^2 and the molar absorbitivity is 6300. The coloured extract is stable for a few days when kept in a cool dark place. After a week, the absorbance changes by only 4%.

Procedure

To ensure that the vanadium in the sample soln. is all present as vanadium(V), treat the soln. (containing ≯ 0·12 mg of vanadium) with a few drops of dilute potassium permanganate soln. until a faint pink colour persists. To this soln. in a separating funnel add the required amounts of hydrochloric acid and water so that the final volume is 25 ml, with an acidity between 2·7 and 7·5 N. Add an 0·1% CPHA soln. in alcohol-free chloroform (8–10 ml), shake, and allow the layers to separate. Collect the chloroform layer in a 50-ml beaker containing anhydrous sodium sulphate (1·5 g). Wash the aqueous layer twice with 5-ml portions of alcohol-free chloroform and add these to the extract in the beaker. Transfer the chloroform soln. from the beaker to the 25-ml flask, wash the sodium sulphate crystals with small portions of chloroform, combine the washings with the soln. in the flask, dilute to the mark and measure its absorbance at 540 nm in a 1-cm cuvette against pure chloroform.

Effect of Other Ions

For the determination of 0·093 mg of vanadium per 25 ml, the following ions, even when present in amounts 250 times that of vanadium, do not interfere: aluminium, barium, calcium, cadmium, cerium(IV), chromium(III), copper, iron, mercury(II), magnesium, manganese, nickel, strontium, thorium, uranium (VI), tungsten(VI), zinc, zirconium, cobalt, acetate, borate, citrate, nitrate, perchlorate, phthalate, phosphate, sulphate and tartrate. The permissible limit for titanium is 20 ppm for each ppm of vanadium(V). Molybdate can be tolerated, if several successive extractions with the chloroformic reagent solution are made for the complete recovery of vanadium.

SUCCESSIVE EXTRACTION AND DETERMINATION OF IRON, VANADIUM AND URANIUM
(Zharovskii and Sukhomlin[37])

CPHA forms complexes with vanadium(V), iron(III) and uranium(VI) which are extractable into organic solvents at different acidities. For instance, the uranium(VI) complex, at pH 5·5–8·5, is extracted quantitatively into ethyl acetate, butanol and iso-amyl alcohol. The maximal absorption of the extract in iso-amyl alcohol is 355 nm, at which wavelength the molar absorptivity (ϵ) is 16,000. The iron(III) complex is extracted into chloroform from solutions of pH 1–4 or into iso-amyl alcohol from solutions of pH 1–7. In the latter solution it shows two absorption maxima, one at 360 nm ($\epsilon = 15,000$) and the other at 480 nm ($\epsilon = 3400$). The condition for vanadium extraction into chloroform from 4 N hydrochloric acid and the measurement of its violet colour compares favourably with that described previously.[36] The mole ratio method indicates that uranium and iron form 1:3 complexes with CPHA.

The extractive photometric method which has been developed is based on the successive extraction of the elements as complexes from mixtures of 0·01–0·1 mg each of vanadium and iron and 0·1–1·0 mg of uranium present in various proportions.

Procedure

Adjust the acidity of the sample soln. to 4 M in hydrochloric acid, and extract vanadium with a 0·1% soln. of CPHA in chloroform. Add ammonia soln. to the aqueous phase to maintain its pH at 1 followed by ethanolic 0·5% soln. of CPHA (2 ml). Extract the iron(III) complex with iso-amyl alcohol (5 ml).

To the aqueous phase, now containing only uranium(VI), add more dilute ammonia soln. to bring the pH to 6–7, add a 0·5% soln. of CPHA in ethanol (2 ml) and extract the uranium complex with iso-amyl alcohol (5 ml).

Measure the absorbances of the extracts of the vanadium(V), iron(III) and uranium(VI) complexes at 536, 453 and 413 nm, respectively, after diluting to 25 ml with the respective solvents used for the extraction.

DETERMINATION OF TITANIUM AND NIOBIUM
(Dutt and Seshadri[38])

Both titanium(IV) and niobium form in 8–9 N hydrochloric acid golden-yellow complexes in the presence of ammonium thiocyanate. The titanium species can also be produced in 10–12 N sulphuric acid. The complexes on extraction into chloroform obey the Lambert–Beer law from 0·25 to 6·00 ppm. of titanium at 410 nm and up to 6·4 ppm of niobium at 420 nm. The sensitivity of the colour reaction for niobium appears to be quite high. The niobium complex, though showing maximal absorption at 370 nm, is always measured at 420 nm, because at the former wavelength the absorbance is too high to measure accurately, even as little as 0·32 ppm of niobium in a 1-cm cell. Both colour systems are stable for 2 hr; the reagent used for colour development is present in an amount 8–10 times the metal ions present.

Procedure

Take an aliquot of either of the metal ion solns. in a separating funnel, add successively hydrochloric acid, to make the 15-ml soln. 8–9 N in acid, ammonium thiocyanate (0·2–0·3 g), 0·5% reagent soln. in ethanol-free chloroform as required and 5 ml of the chloroform. Shake for a few min to extract the golden-yellow complex. Repeat the extraction with 3-ml portions of ethanol-free chloroform till the aqueous layer becomes colourless. Five such extractions are usually required. Combine the extracts, dry over fused calcium chloride, filter through a dry filter paper into a 25-ml graduated flask, make up to volume with chloroform and measure its absorbance at 420 nm in a 1-cm cell against a reference solution.

Effect of Other Ions

During the measurement of the colour intensity of the titanium species, it has been observed that oxalic, tartaric and phosphoric acids and aluminium do not interfere. Vanadium(V), molybdenum(VI) and tungsten(VI) up to 2–3 times and tin(IV) and zirconium(IV) up to 8–10 times the amount of titanium also do not interfere. Fluoride interferes, as does copper(II) unless masked by thiourea.

For niobium determinations, zirconium, vanadium(V), molybdenum(VI) and tungsten(VI) up to 10 times, aluminium in large amounts, oxalate and fluoride, 2–3 times, and sulphate up to 10–11 times the amount of niobium can be tolerated. Tantalum and titanium interfere seriously.

6. N-Benzoyl-p-tolylhydroxylamine as a Spectrophotometric Reagent

DETERMINATION OF TITANIUM
(Dutt and Seshadri[38])

The golden-yellow complex of titanium(IV) formed in 8–9 N hydrochloric or 10–12 N sulphuric acid in the presence of ammonium thiocyanate, on extraction into chloroform shows maximal absorption at 370 nm. At this wavelength the system obeys the Lambert–Beer law from 0·25 to 6·00 ppm of titanium(IV). As the reagent is colourless a large excess has no effect on the colour system, which is stable for 2 hr.

Procedure

Follow that described on p. 140 and measure the absorbance at 370 nm against the reference solution.

Effect of Other Ions

These are as stated above for titanium species (p. 141).

7. N-Benzoyl-o-tolylhydroxylamine as a Spectrophotometric Reagent

DETERMINATION OF VANADIUM
(Majumdar and Das[39])

The reddish-violet complex of vanadium(V) with N-benzoyl-o-tolylhydroxylamine (NBTHA) on extraction into chloroform has an absorbance at maximum 510 nm. The acid concentration for maximal absorbance is from 4 to 8 N hydrochloric acid. The Lambert–Beer law is obeyed for 0·5–10 μg of vanadium, but the optimal concentration range is 2–10 μg per ml of chloroform. The colour system is stable for about 5 days and a large excess of the reagent can be used even for very small amounts of vanadium because the excess of reagent has no effect on the colour. The composition of the complex as determined by a Job plot and by mole ratio and slope ratio methods confirms the presence of the metal and the reagent in the complex in a ratio of 1:2, the instability constant of which is in the order of 10^{-9}. The sensitivity calculated is 0·0108 μg of vanadium per cm^2 and the molar absorptivity is 4743. The procedure is specific for the spectrophotometric determination of vanadium if fluoride is present.

Reagents

Vanadium solution. Prepare an aqueous soln. of ammonium metavanadate with the addition of a few drops of dilute ammonia. Determine the vanadium concentration by an EDTA titration.

Alcohol-free chloroform. Wash chloroform successively with (1 + 1) sulphuric acid, dilute ammonia soln. and distilled water. Dry over fused calcium chloride and distil.

NBTHA solution. 0·5% in alcohol-free chloroform.

Procedure

Take an aliquot of the vanadate soln. (containing 0·05–0·25 mg of vanadium) in a separating funnel, add 6 N hydrochloric acid (10 ml) to maintain the acidity at the optical value, followed by the NBTHA soln. (4 ml) and more alcohol-free chloroform (4 ml; 1 ml of NBTHA soln. is sufficient to develop maximal colour with 0·1 mg of vanadium in 25 ml of chloroform). Shake the mixture for a few min and transfer the reddish-violet chloroform layer to a 25-ml graduated flask. Extract any vanadium left in the aqueous phase with a further 5-ml portion of alcohol-free chloroform. Dilute the combined extracts to the mark with the chloroform. Measure the absorbance of the solution at 510 nm in a 1-cm cuvette against a chloroform blank.

Effect of Other Ions

For the determination of vanadium (4 μg per ml) the following ions can be tolerated (amounts in mg, in parenthesis): aluminium (26), iron(III) (20), manganese (20), uranium(VI) (20), chromium(III) (15), zirconium (20), copper (20), molybdenum(VI) (20), titanium (15), tungsten(VI) (15), thorium (20), cobalt (15), cadmium (20), mercury(II) (20), lead (15), arsenic(III) (10), arsenic(V) (10), beryllium (20), magnesium (20), barium (20), tin(IV) (20), bismuth (20), nickel (15), zinc (15), calcium (15), strontium (15), rare earths (10), hydrogen peroxide (2 ml of 20 vol.), citrate (10), oxalate (15), tartrate (15), EDTA (20), phosphate (20), fluoride (20). Titanium(IV) must be masked by fluoride.

8. *N*-Benzoyl-*p*-chlorophenylhydroxylamine as a Spectrophotometric Reagent

DETERMINATION OF VANADIUM
(Majumdar and Das[40])

Vanadate forms with *N*-benzoyl-*p*-chlorophenylhydroxylamine (NBCHA) in strong hydrochloric acid a violet complex which is extractable into chloroform. This complex has maximal absorption at 530 nm. Maximal colour development occurs in 4–8 N hydrochloric acid. At 530 nm, the Lambert–Beer law holds for 0·5 to 8 μg of vanadium, and the optimal concentration range is from 2 to 8 μg of vanadium per ml of chloroform. The colour intensity is constant at room temperature for more than 5 days. The reagent has no deleterious effect on the colour system, so a large excess can be used. The determination of the composition of the complex by Job's method of continuous variation and the mole ratio method suggests that the complex contains the metal and the reagent at a ratio of 1:2. The dissociation constant of the complex, as determined by these methods is 10^{-8}. The molar absorptivity is 4500. The sensitivity of the reaction at 530 nm is 0·011 μg of vanadium per cm^2. The vanadium complex is also soluble in benzene, xylene, carbon tetrachloride and methyl isobutyl ketone; the wavelength of maximal absorption of the complex is the same in all these solvents. The presence of ethanol or acetone reduces the colour intensity.

Procedure

To the vanadate soln. (containing 0·05–0·2 mg of vanadium) in a separating funnel, add hydrochloric acid (10 ml) to adjust the acidity to 6 N, followed successively by an 0·5% NBCHA soln. in alcohol-free chloroform (4 ml) and alcohol-free chloroform (6 ml; see p. 142). Shake for a few min and allow the chloroform layer to separate. To ensure complete recovery of vanadium, shake the aqueous layer once more with alcohol-free chloroform (5 ml). Combine the extracts in a 25-ml graduated flask and make up to volume with the chloroform. Measure the absorbance at 530 nm in a 1-cm cuvette against chloroform.

Effect of Other Ions

Vanadium (4 μg per ml) can be determined in presence of the following ions (mg in parenthesis): aluminium (20), barium (20), calcium (15), cobalt (15), chromium(III) (20), copper (20), iron(III) (30), magnesium (15), titanium (30), manganese (20), nickel (20), thorium (24), tin(IV) (18), uranium(VI) (20), zinc (10), zirconium (20), beryllium (20), cadmium (15), mercury(II) (15), lead (20), arsenic(III) (10), arsenic(V) (15), rare earths (20), bismuth(III) (20), acetate (15), borate (15), citrate (10), tartrate (15), oxalate (20), phosphate (20), fluoride (30), EDTA (20). Copper(II) and titanium(IV) must be masked, however, with EDTA and fluoride, respectively.

9. N-Benzoyl-o-Tolylhydroxylamine and Other Aromatic Hydroxylamines as Spectrophotometric Reagents

DETERMINATION OF VANADIUM
(Majumdar and Das[41] and Das[42])

N-Benzoyl-o-tolylhydroxylamine (I), as noted above (p. 142), is a specific spectrophotometric reagent for vanadium, while N-benzoyl-p-chlorophenylhydroxylamine (II) (see p. 143) is highly selective in its reaction towards the element. Other derivatives such as N-phenylacetylphenylhydroxylamine (III), N-benzoyl-m-tolylhydroxylamine (IV) and N-benzoyl-p-tolylhydroxylamine (V) have reaction characteristics similar to the o-tolyl derivative (cf. Table 5.1). With vanadate, (III) forms a reddish-violet complex, and (IV) and (V) give violet complexes. These coloured complexes, on extraction into chloroform from strong hydrochloric acid, form the basis of spectrophotometric methods for the determination of vanadium. However, in the presence of even traces of ethanol the colour intensities are

TABLE 5.1. REACTIONS OF SOME AROMATIC HYDROXYLAMINES WITH VANADIUM(V) IN HYDROCHLORIC ACID

	N-Benzoyl-o-tolyl-hydroxylamine	N-Benzoyl-m-tolyl-hydroxylamine	N-Benzoyl-p-tolyl-hydroxylamine	N-Phenyl-acetylpheny-hydroxylamine
Wavelength of maximal absorbance (nm)	510	530	530	510
Lambert–Beer law range (μg/ml)	0·5–10·0	0·5–10·0	0·5–10·0	0·5–10·0
Optimal range (μg/ml)	2–10	2–10	2–10	2–8
Optimal hydrochloric acid concentration (N)	4·0– 8·0	2·7– 8·0	2·7– 7·5	2·8– 7·0
Composition of complex (metal:reagent)	1:2	1:2	1:2	1:2
Average dissociation constant at room temperature	8×10^{-9}	$6·4 \times 10^{-9}$	$4·0 \cdot 10^{-9}$	$3·8 \times 10^{-9}$
Sensitivity (μg/cm^2)	0·0108	0·0108	0·0108	0·0121
% Relative error, in the optimal vanadium concentration range	2·73	2·73	2·72	2·72

TABLE 5.2. REACTIONS OF SOME AROMATIC HYDROXYLAMINES WITH VANADIUM(V) AT pH 4·8–6·0

	N-Benzoyl-o-tolyl-hydroxylamine	N-Benzoyl-m-tolyl-hydroxylamine	N-Benzoyl-p-tolyl-hydroxylamine	N-Phenylacetyl-phenyl-hydroxylamine	N-Benzoyl-p-chlorophenyl-hydroxylamine
Composition of complex (metal:reagent)	1:2	1:2	1:2	1:2	1:2
Average dissociation constant at room temperature	2.4×10^{-8}	2.3×10^{-8}	2.5×10^{-8}	2.9×10^{-8}	3.1×10^{-8}
Sensitivity (μg/cm^2)	0·0129	0·0129	0·0129	0·0145	0·0137
% Relative error, at the optimal of vanadium concentration range (3–12 μg)	2·74	2·74	2·74	2·72	2·73

greatly reduced and the colours are different. Thus the violet solutions of the vanadium(V) complexes of all the five reagents become orange on the addition of ethanol and both the violet and the orange complexes are extractable into solvents like benzene, xylene, chloroform and carbon tetrachloride. Therefore, with chloroform as the solvent, two extraction processes have been developed, one from hydrochloric acid and the other from an aqueous ethanolic solution, pH 4·8–6·0. In these two processes, a marked difference in the number of interfering ions is observed.

The yellow complexes formed at pH 4·8–6·0 (sodium acetate–acetic acid buffer) on extraction into chloroform have absorption maxima at 440 nm and obey the Lambert–Beer law for 1–12 μg of vanadium with the optimal range from 3–12 μg of vanadium per ml. By the continuous variation, mole ratio and slope ratio methods, the composition of the complexes appears to be 1:2 (metal:reagent) with dissociation constants of the order of 10^{-8}. Under the experimental conditions, the tolyl derivatives are more sensitive (cf. Table 5.2) and for 8 μg of vanadium, only 2 ml of a 0·5% reagent solution in ethanol is sufficient to give maximal colour intensity, which is stable for 48 hours.

Reagents

Aromatic hydroxylamine solutions. 0·5% solns. of the substituted hydroxylamines in alcohol-free chloroform (p. 142) and also in ethanol.

Buffer solution. Prepare solns. of pH 4·8–6·0, by mixing suitable amounts of 0·2 M sodium acetate soln. with 0·2 M acetic acid.

Procedure (a)

Add hydrochloric acid (10 ml) to the vanadate solution to make it 6 N in acid. Add the substituted hydroxylamine soln. (4 ml) and the alcohol-free chloroform (4 ml). Shake for a few min and collect the extract in a 25-ml graduated flask. Add more purified chloroform (5 ml) to the aqueous layer, shake and combine this extract with the previous one. Make up the volume of the combined extracts to 25 ml and measure the absorbance in a 1-cm cuvette at the region of its maximum absorbance as given in Table 5.1, against chloroform.

Effect of Other Ions

Vanadium(V) (4 μg per ml) can be determined by extraction with a chloroform solution of any of the reagents (I), (III), (IV) and (V), in the presence of the following ions (amount in mg in parenthesis): aluminium (20), iron(III) (30), manganese (20), uranium(VI) (20), chromium(III) (20), zirconium (20), copper (20), molybdenum(VI) (20), titanium (10), tungsten(VI) (15), thorium (20), cobalt (15), cadmium (15), mercury(II) (15), lead (20), arsenic(III) (10), arsenic(V) (15), beryllium (20), magnesium (15), barium (20), tin(IV) (15), bismuth (20), nickel (20), zinc (10), antimony (15), calcium (15), strontium (10), lanthanides (20), osmium(VIII) (0·012), iridium(IV) (1·8), rhodium(III) (0·027), ruthenium(III) (0·025), palladium (1·22), platinum(IV) (2·3), EDTA (20), fluoride (10), oxalate (15), tartrate (15), phosphate (20), citrate (10), acetate (15), borate (10). Titanium must be masked by fluoride. Although the presence of 2 ml of 20 vol. hydrogen peroxide has no effect when reagent (I) is used, the presence of only 0·2 ml requires several extractions for the complete recovery of vanadium if one of the other three reagents is used. Similarly, if molybdate is present, five extractions by reagent (V) or eight extractions by reagent (III) are required.

Procedure (b)

To the faintly ammoniacal vanadium(V) soln. (containing 0·075–0·30 mg of vanadium) add the sodium acetate–acetic acid buffer soln. (15 ml) to maintain the pH at about 5·2, followed by one, of the 0·5% reagent solns. in ethanol (4 ml). Shake twice with 5-ml portions of chloroform completely to extract the vanadium complex. Transfer the extracts to a 25-ml graduated flask, add more chloroform to make up to volume and measure the absorbance at 440 nm in a 1-cm cuvette against chloroform.

Effect of Other Ions

At pH 4·8–6·0, 8 μg of vanadium per ml can be determined in the presence of a large excess of a limited number of ions (amounts in mg in parenthesis): barium (15), calcium (15), magnesium (15), strontium (15), beryllium (20), chromium(III) (20), manganese (10), cobalt (15), nickel (15), zinc (10), cadmium (20), mercury(II) (10), arsenic(III) (15), arsenic(V) (15), fluoride (15), tartrate (10), borate (15), phosphate (15).

DETERMINATION OF VANADIUM IN SILICATE ROCKS AND MINERALS WITH N-BENZOYL-o-TOLYLHYDROXYLAMINE*
(Jeffery and Kerr[43])

In this investigation, practically the same method as recommended previously[39–42] for the extraction of the vanadium(V)-N-benzoyl-o-tolylhydroxylamine complex from a 4–8 N hydrochloric acid and the determination of vanadium in the extract at 510 nm has been followed, but with the use of carbon tetrachloride as the extractant, in place of alcohol-free chloroform, to avoid the interference arising out

* The information is Crown Copyright.

of the possible presence of alcohol. The complex formed between vanadium(V) and *N*-benzoyl-*o*-tolylhydroxylamine (NBTHA) is soluble in several organic solvents such as carbon tetrachloride, chloroform, isobutyl methyl ketone and toluene. The calibration graph is the same for all the solvents.

For the determination of vanadium in rocks and minerals, the residue after the removal of silica is fused with potassium pyrosulphate, dissolved in dilute sulphuric acid and oxidized with permanganate. Vanadium(V) present in the solution is then extracted from 6 N hydrochloric acid by a carbon tetrachloride solution of NBTHA and the absorbance of the extract is measured at 510 nm. Titanium must be masked by fluoride.[39]

Procedure

Heat the finely powdered silicate-containing sample (0·1 g) in the usual way with sulphuric, nitric and hydrofluoric acids. Remove sulphuric acid by heating on a hot plate. Fuse the residue in a silica crucible with potassium pyrosulphate and extract the melt with water (10 ml) containing 20 N sulphuric acid (2 drops). Transfer to a separating funnel, add 0·02 M potassium permanganate soln. dropwise until the soln. has a pink colour which lasts for 5 min. Dilute to 20 ml, add an 0·05 M sulphamic acid soln. (2 ml), a saturated sodium fluoride soln. (2 ml), conc. hydrochloric acid (20 ml) and an 0·02% NBTHA soln. in carbon tetrachloride (10 ml). Shake for 30 sec to extract the vanadium complex. Filter the lower organic layer through a wad of cotton wool and measure its absorbance at 510 nm in a 2-cm cuvette, against a reagent blank.

To prepare calibration graphs, take standard vanadate solutions (containing 10–50 μg of vanadium) in a number of separating funnels, dilute each to 20 ml, add potassium permanganate to each in slight excess and determine vanadium as described above.

DETERMINATION OF VANADIUM
(Majumdar and Bhowal[44])

In a recent investigation, ethanol-free chloroform is found to be a better extractant than carbon tetrachloride for the vanadium–NBTHA complex. Both the reagent and its vanadium complex are highly soluble in chloroform so that only two extractions are sufficient to completely extract the vanadium complex into chloroform, whereas at least four extractions are needed for the complete removal of the complex into carbon tetrachloride. The colour reaction also is more sensitive in chloroform than in carbon tetrachloride; the spectrophotometric sensitivity values being 0·0108 and 0·0123 μg of vanadium per cm^2, respectively. The absorbances are the same whether aliquots of standard vanadate solutions are oxidized or not before treatment with the reagent solution.

The presence of a very small proportion of vanadium(IV) is not expected to affect the result as it is observed that when a 4–8 N hydrochloric acid solution of vanadium(IV) is kept in contact with a chloroformic reagent solution for some time and shaken, the chloroform extract assumes the red-violet colour of the vanadium(V) complex, with the same maximal absorption at 510 nm. Moreover, the solid complex isolated on the addition of an acetone solution of the reagent to the hydrochloric acid solution of vanadium(IV) appears to be identical with the violet complex of vanadium(V). Both complexes melt at 133° and are diamagnetic. From this it can be inferred that vanadium(V) forms the more stable complex with the reagent with the result that the reduction potential of the vanadium (IV)/vanadium(V) couple increases and hence vanadium(IV) is quickly oxidized to vanadium(V) by air to produce the violet complex.

10. N-Furoylphenylhydroxylamine as a Spectrophotometric Reagent

DETERMINATION OF TITANIUM IN STEELS, ALUMINIUM ALLOYS AND ORES
(Pilipenko, Shpak and Boiko[45])

Since N-furoylphenylhydroxylamine (NFPHA) gives chloroform-extractable coloured complexes with titanium, iron(III) and vanadium (V), the latter two are reduced by tin(II) chloride to the di- and tetra-valent states, respectively, before the determination of titanium.

Addition of ammonium thiocyanate considerably increases the sensitivity of the titanium reaction in sulphuric acid.

Procedure

Dissolve the steel sample (0·10–0·25 g, containing 10–50 μg of titanium) in 10 ml of hydrochloric acid, add 30% hydrogen peroxide (1 ml) and boil for 5 min to decompose the carbides. Cool, and dilute to 25 ml with hydrochloric acid. Take a 10 ml aliquot, add to it a 10% tin(II) chloride soln. in hydrochloric acid (1 ml) and a 2% soln. of NFPHA in methanol (1·5 ml).

Dilute to 25 ml with hydrochloric acid and extract the titanium complex with chloroform (5 ml). Add more reagent soln. (0·5 ml) to the aqueous phase and extract with another portion (5 ml) of chloroform. Combine the extracts, dilute with chloroform to 10 ml and measure the absorbance at 413 nm against chloroform in a 1-cm cuvette.

For the determination in aluminium alloys, dissolve the sample (0·5–1·0 g, containing 10–50 µg of titanium) in hydrochloric acid (5 ml) and dilute to 25 ml with the acid. To a 10-ml aliquot, add in succession a 1% tin(II) chloride soln. (0·5 ml), reagent soln. (1·5 ml), and dilute to 25 ml with hydrochloric acid. Then follow the procedure described for the determination of titanium in steels.

For the determination in ores, dissolve by decomposing the sample (0·1 g, containing 10–80 µg of titanium) in a mixture of hydrofluoric and sulphuric acids. Evaporate to remove hydrofluoric acid and most of the sulphuric acid and dilute to 25 ml with (2 + 3) sulphuric acid. Take an aliquot (5–10 ml), add to it a 10% ammonium thiocyanate soln. (1·5 ml) and reagent soln. (1·5 ml). Dilute to 25 ml with dilute sulphuric acid and extract with chloroform (5 ml).

To the aqueous phase, add 0·5 ml each of the thiocyanate and reagent solns., extract again with chloroform (5 ml) and measure the absorbance in the total extract as described above.

DETERMINATION OF VANADIUM IN BRINE
(Pilipenko, Sereda and Shpak[46])

N-Furoylphenylhydroxylamine (NFPHA) is more sensitive to vanadium than N-benzoylphenylhydroxylamine. It has been used for the determination of µg amounts of vanadium in brine. The coloured complex on extraction into chloroform obeys the Lambert–Beer law for 1–12 µg of vanadium.

Procedure

To determine 10^{-5}% of vanadium in brine, dilute the brine (containing 20 g of sodium chloride) to 80 ml. Mix with a 1 M sodium or ammonium fluoride soln. (1 ml) to mask interfering ions. Add conc. hydrochloric acid (20 ml). Shake for 1 min with an 0·5% soln. of the reagent in chloroform (5 ml), filter the extract through a dry filter paper and measure its absorbance at 510 nm in a 1-cm cell against the reagent soln. as blank.

DETERMINATION OF VANADIUM IN STEELS AND ORES
(Pilipenko, Shpak and Kurbatova[47])

N-Furoylphenylhydroxylamine (RH) forms with vanadium(V) a brown complex in a weakly acidic medium and a violet complex in a strongly acidic medium. The complexes are readily extracted into chloroform and other organic solvents. A small amount of ethanol has a favourable effect on the extraction of vanadium into the organic layer, but a large amount of ethanol in chloroform decreases the absorbance of the extract. Because during the extraction from 7 N hydrochloric acid, vanadium(V) is partially reduced to vanadium (IV), chloroform extraction must be carried out in 2 min.

The coloured system formed on the extraction of the vanadium(V) complex by chloroform from a strong hydrochloric acid solution has maximal absorption at 530 nm with a molar absorptivity of 5650. The complex extracted from a weakly acidic solution shows maximal absorption at 450 nm with a molar absorptivity of 1800.

For the extraction and photometric determination of μg amounts of vanadium, the optimal acidities are 3–7 N hydrochloric acid, 7–13 N sulphuric acid and pH 2–4. The sensitivity, however, is greatest in hydrochloric acid. On extraction from 6 N hydrochloric acid, the system obeys the Lambert–Beer law over the range 0·5–5·0 μg of vanadium per ml. The use of ammonium fluoride to mask the interference of titanium, niobium, molybdenum and tungsten has been recommended.

The complex isolated from 4 N hydrochloric acid is VOR_2Cl and that from a solution at pH 4 is $VO_2R \cdot RH$; after extraction into chloroform the complexes are VOR_2Cl and VO_2R, respectively.

Procedure

Weigh out the sample (0·05–0·20 g, containing 25–250 μg of vanadium) and decompose it with (1 + 4) sulphuric acid (40 ml). After the sample is entirely dissolved, add perhydrol dropwise until the soln. is decolorized. Heat until the carbide particles disappear. Cool, transfer to a 50-ml graduated flask using (1 + 4) sulphuric acid (10 ml) as wash liquid and make up to volume with water.

Alternatively, dissolve the sample (0·1 g, containing 25–250 μg of vanadium) in a mixture of hydrofluoric and nitric acids. Add conc.

sulphuric acid (10 ml). Heat to fumes. Cool, dilute with water, transfer to a 50-ml graduated flask with (1 + 4) sulphuric acid (10 ml) as wash liquid and make up to volume with water.

To a 10-ml aliquot of this soln., add dropwise 0·1 N potassium permanganate soln. until the soln. becomes pink. Wait for 1 min. Dilute to 25 ml with (1 + 1) hydrochloric acid and extract vanadium with an 0·2% NFPHA soln. in chloroform (10 ml). Measure the absorbance at 536 nm in a 1-cm cuvette against the chloroform extract of a soln. treated as above but containing no vanadium.

11. N-Acetylsalicylphenylhydroxylamine as a Spectrophotometric Reagent

IDENTIFICATION AND DETERMINATION OF TITANIUM
(Savariar and Joseph[48])

N-Acetylsalicylphenylhydroxylamine forms with titanium(IV) in strong hydrochloric acid a yellow, chloroform-extractable complex. The selectivity and the sensitivity of the colour reaction in the presence of a large excess of thiocyanate increase so considerably that a highly selective spot test for titanium, with an identification limit of 0·1 μg and a dilution limit of $1:2 \times 10^6$ and a spectrophotometric method for titanium have been developed, based on this enhancement.

For maximal colour development in the extract from 4·5–8·0 M hydrochloric acid containing 12·5–88 μg titanium, the optimal amount of reagent is 3 ml of an 0·5% solution in chloroform and that of the thiocyanate is 3 ml of a 4 M solution in water. The coloured extract, which is stable for 24 hr at room temperature and is unaffected by variation of temperature from 15° to 55°, has no well-defined absorption maximum. However, its absorption increases as the wavelength decreases from 480 nm and so spectrophotometric measurements are made at 390 nm, where the reagent absorption is negligible. The extract at this wavelength obeys the Lambert–Beer law from 0·25 to 4·00 ppm of titanium with the optimal range from 0·5 to 3·5 ppm, where the per cent relative error is only 1·85. The photometric sensitivity and the molar absorptivity are, respectively, 0·004 μg of titanium per cm^2 and 10,900. In the extract, titanium and the reagent combine in a ratio of 1:2.

Spot Test for Titanium

Take, in a micro test tube the titanium soln. (0·05 ml), 4 M ammonium thiocyanate soln. (0·05 ml) and 10 M hydrochloric acid (0·1 ml). Shake for 1 min with an alcohol-free chloroformic 0·5% reagent soln. (0·2 ml). A deep-yellow chloroform layer indicates the presence of titanium (IV).

Effect of Other Ions

Even 1000-fold excesses of thallium(I), zinc, cadmium, mercury(II), magnesium, beryllium, iron(II and III), lead, manganese, calcium, strontium, barium, arsenic(III), antimony, bismuth, aluminium, chromium, lanthanum, cerium(IV), zirconium, thorium, tin(IV), vanadium(V), tantalum, molybdenum(VI), tungsten(VI), uranium(VI), tellurium(VI) and osmium(VIII) do not interfere. Copper (II) and niobium, masked, respectively, by EDTA and oxalate, are tolerated in 1000-fold amounts. Of the anions, tartrate, citrate, borate, oxalate, persulphate, phosphate and EDTA do not interfere even when present in more than 1000-fold excesses. Fluoride interferes unless masked by boric acid.

Determination of Titanium

To an aliquot of the soln. (containing 12·5–88·0 μg of titanium) in a separating funnel, add a 4 M ammonium thiocyanate soln. (2–4 ml) followed by 10 M hydrochloric acid to make the soln. 4·5–8·0 M in acid, and an 0·5% soln. of reagent in alcohol-free chloroform (3–4 ml). Shake to extract for 2 min with alcohol-free chloroform (5 ml). Transfer the chloroform layer to a small beaker containing a few g of anhydrous sodium sulphate. Extract again with another portion of alcohol-free chloroform (5 ml) and collect the extract in the same beaker. Decant the entire extract into a 25-ml graduated flask, wash the beaker with a few ml more of the chloroform, transfer the washings to the flask, make up to volume with chloroform and measure its absorbance at 390 nm against a reagent blank.

Effect of other Ions

This is the same as is described for the spot test.

12. N-Arylhydroxylamines as Spectrophotometric Reagents

DETERMINATION OF VANADIUM(V)
(Tandon and Tandon[49])

The vanadium(V) complexes of the N-arylhydroxylamines formed in strong hydrochloric acid media can be extracted by solvents like benzene, diethyl ether, o-dichlorobenzene, ethyl acetate, carbon tetrachloride and chloroform, although chloroform appears to be the most suitable extractant. The coloured complexes extracted into chloroform from 4 N hydrochloric acid show regions of maximal absorption from 500 to 540 nm with molar absorptivities of 3850 to 6100 (Table 5.3); the wavelength of maximal absorption is constant for hydrochloric acid concentrations above 2 N. The composition of the complexes extracted both from strongly acidic (~ 4 N) and weakly acidic (pH 1–6) solutions, as determined by Job's method, always indicates a metal to reagent ratio of 1:2.

TABLE 5.3. COLOUR, ABSORPTION MAXIMUM AND MOLAR ABSORPTIVITY OF VANADIUM(V)–N-ARYLHYDROXYLAMINE COMPLEXES IN CHLOROFORM

Hydroxylamines	Colour	λ_{max} nm	ϵ
N-p-Methylbenzoyl-p-tolyl-	Violet	530	4900
N-p-Fluorobenzoylphenyl-	,,	525	4400
N-p-Fluorobenzoyl-p-tolyl-	,,	530	4500
N-m-Bromobenzoyl-p-tolyl-	,,	530	4450
N-p-Bromobenzoyl-p-tolyl-	,,	530	4650
N-p-Bromobenzoylphenyl-	,,	525	4550
N-m-Methylbenzoylphenyl-	Blue-violet	530	5150
N-o-Methylbenzoyl-p-tolyl-	,,	530	4750
N-m-Methylbenzoyl-p-tolyl-	,,	535	5250
N-o-Bromobenzoylphenyl-	,,	530	4450
N-o-Bromobenzoyl-p-tolyl-	,,	535	4500
N-o-Chlorobenzoyl-p-tolyl-	,,	530	4500
N-o-Iodobenzoyl-p-tolyl-	,,	535	4400
N-p-Methoxybenzoylphenyl-	,,	535	5900
N-p-Methoxybenzoyl-p-tolyl-	,,	540	6100
N-Phenylacetyl-p-tolyl-	Red-violet	505	4500
N-Phenoxyacetylphenyl-	,,	500	3850
N-p-Chlorophenoxyacetylphenyl-	,,	500	3850
N-n-Butyrylphenyl-	,,	500	4400
N-n-Butyryl-p-tolyl-	,,	505	4500
N-Laurylphenyl-	,,	500	4400
N-Lauryl-p-tolyl-	,,	505	4500
N-Palmitylphenyl-	,,	500	4400
N-Palmityl-p-tolyl-	,,	505	4500

Procedure

Oxidize vanadium in the sample soln. by the dropwise addition of dilute potassium permanganate soln. Take an aliquot (containing 0·02–0·2 mg of vanadium(V)) in a separating funnel, dilute with water and hydrochloric acid so that the volume is 25 ml with an acidity between 2 and 10 N. Add a 0·005 M chloroform soln. of an N-arylhydroxylamine (10 ml) and follow the method as described on p. 139 for the extraction and the measurement of the absorbance in a 1-cm cell against chloroform, at the wavelength given in Table 5.3.

References

1. SANDELL, E. B., *Colorimetric Determination of Traces of Metals*, 3rd ed. Interscience Publishers, New York, 1959.
2. RINGBOM, A., *Z. Anal. Chem.* **115**, 332 (1938/9).
3. AYRES, G. H., *Anal. Chem.* **21**, 652 (1949).
4. JOB, P., *Compt. Rend.* **180**, 928 (1925); *Ann. Chim. (Paris)*, **9**, 113 (1928).
5. YOE, J. H. and JONES, A. L., *Ind. Eng. Chem., Anal. Ed.* **16**, 111 (1944).
6. HARVEY, A. E. and MANNING, D. L., *J. Am. Chem. Soc.* **72**, 4488 (1950).
7. MAJUMDAR, A. K. and SEN, B., *Anal. Chim. Acta* **8**, 369 (1953).
8. SHOME, S. C., *Anal. Chem.* **23**, 1186 (1951).
9. ZHAROVSKII, F. G. and PILIPENKO, A. T., *Ukrain. Khim. Zhur.* **25**, 230 (1959).
10. RYAN, D. E., *Analyst* **85**, 569 (1960).
11. PRIYADARSHINI, U. and TANDON, S. G., *Anal. Chem.* **33**, 435 (1961).
12. DE POOL, D. H. and CADAVIECO, R. D., *Acta Cient. Venezolana* **13**, 157 (1962); *Anal. Abstr.* **10**, 3185 (1963).
13. TOMIOKA, H., *Bunseki Kagaku* **12**, 271 (1963); *C.A.* **59**, 5768 (1963).
14. ANTONIJEVIC, V., *Glas. Hem. Drus. Beograd* **31**, 305 (1966); *C.A.* **70**, 74003 (1969).
15. PATROVSKY, V., *Chem. Listy* **60**, 1545 (1966); *Anal. Abstr.* **15**, 739 (1968).
16. BAUGHMAN, W. J. and WATERBURY, G. R., *U.S. At. Energy Comm.* 1968, *LA-3843*, 9 pp.
17. IWASAKI, I., OZAWA, T. and YOSHIDA, S., *Bunseki Kagaku* **17**, 986 (1968); *C.A.* **70**, 43843 (1969).
18. PILKINGTON, E. S. and WILSON, W., *Anal. Chim. Acta* **47**, 461 (1969).
19. HOFER, A. and HEIDINGER, R., *Z. Anal. Chem.* **246**, 125 (1969).
20. YUN-HSIANG, CHIEH, Candidate's Thesis, Moscow State University, 1960; *Uspekhi Khim* **31**, 989 (1962).
21. ISHII, H. and EINAGA, H., *Bunseki Kagaku* **17**, 1296 (1968); *C.A.* **70**, 53675 (1969).
22. ZHAROVSKII, F. G., SHPAK, E. A. and PISKUNOVA, E. V., *Ukrain. Khim. Zhur.* **28**, 1104 (1962).
23. SCHWARBERG, J. E. and MOSHIER, R. W., *Anal. Chem.* **34**, 525 (1962).
24. TANAKA, K. and TAKAGI, N., *Bunseki Kagaku* **12**, 1175 (1963).
25. VITA, O. A., MULLINS, L. R. Jr. and TRIVISONNO, C. F., GAT-T-1085 (1963).
26. AFGHAN, B. K., MARRYATT, R. G. and RYAN, D. E., *Anal. Chim. Acta* **41**, 131 (1968).
27. CHE-MING, NI and SHU-CHUAN, LIANG, *Scientia Sinica* **12**, 615 (1963).
28. CHE-MING, NI and SHU-CHUAN, LIANG, *Hua Hsueh Hsueh Pao* **30**, 540 (1964); *C.A.* **62**, 11142 (1965).
29. VILLARREAL, R. and BARKER, S. A., *Anal. Chem.* **41**, 611 (1969).

30. SHIGEMATSU, T., NISHIKAWA, Y., GODA, S. and HIRAYAMA, H., *Bull. Inst. Chem. Res. Kyoto Univ.* **43**, 347 (1965); *C.A.* **65**, 2982 (1966).
31. DAS, B. and SHOME, S. C., *Anal. Chim. Acta* **35**, 345 (1966).
32. MURUGAIYAN, P. and SANKAR DAS, M., *Anal. Chim. Acta* **48**, 155 (1969).
33. GUPTA, H. K. L. and SOGANI, N. C., *J. Indian Chem. Soc.* **37**, 769 (1960).
34. GUPTA, H. K. L. and SOGANI, N. C., *J. Indian Chem. Soc.* **40**, 15 (1963).
35. TANDON, S. G. and BHATTACHARYYA, S. C., *Anal. Chem.* **33**, 1267 (1961).
36. PRIYADARSHINI, U. and TANDON, S. G., *Analyst* **86**, 544 (1961).
37. ZHAROVSKII, F. G. and SUKHOMLIN, R. I., *Zhur. Anal. Khim.* **21**, 59 (1966).
38. DUTT, N. K. and SESHADRI, T., *Indian J. Chem.* **6**, 741 (1968).
39. MAJUMDAR, A. K. and DAS, GAYATRI, *Anal. Chim. Acta* **31**, 147 (1964).
40. MAJUMDAR, A. K. and DAS, GAYATRI, *J. Indian Chem. Soc.* **42**, 189 (1965).
41. MAJUMDAR, A. K. and DAS, GAYATRI, *Anal. Chim. Acta* **36**, 454 (1966).
42. DAS, GAYATRI, Ph.D. Thesis, Jadavpur University, 1967.
43. JEFFERY, P. G. and KERR, G. O., *Analyst* **92**, 763 (1967).
44. MAJUMDAR, A. K. and BHOWAL, S. K., *Analyst* **96**, 127 (1971).
45. PILIPENKO, A. T., SHPAK, E. A. and BOIKO, YU. P., *Zavod. Lab.* **31**, 151 (1965); *C.A.* **62**, 12438 (1965).
46. PILIPENKO, A. T., SEREDA, I. P. and SHPAK, E. A., *Zavod. Lab.* **32**, 660 (1966); *Anal. Abstr.* **14**, 5370 (1967).
47. PILIPENKO, A. T., SHPAK, E. A. and KURBATOVA, G. T., *Zhur. Anal. Khim.* **22**, 1014 (1967).
48. SAVARIAR, C. P. and JOSEPH, J., *Anal. Chim. Acta* **47**, 347 (1969).
49. TANDON, U. and TANDON, S. G., *J. Indian Chem. Soc.* **46**, 983 (1969).

CHAPTER 6

SEPARATION OF ELEMENTS BY SOLVENT EXTRACTION WITH N-BENZOYLPHENYLHYDROXYLAMINE

N-BENZOYLPHENYLHYDROXYLAMINE (BPHA) was first synthesized in 1919. The solubility of its metal complexes was first investigated in 1944 and up to 1958 studies of the extraction of BPHA complexes was limited to only a very few elements. Today quite a large number (about two-thirds of the elements in the periodic table) have been studied for their extraction characteristics and some novel separations have been achieved.

The extraction of metal ions with a solution of BPHA as extractant is a function of time, concentration of BPHA, nature of the solvent, type of the mineral acid and the acidity of the medium. Solvents used for extractions are usually chloroform, benzene and iso-amyl alcohol, chloroform being the preferred solvent.

For extraction from sulphuric acid, a higher concentration of BPHA is required than for extraction from hydrochloric acid, perhaps because of the formation of the more stable sulphato complexes with the metal ions. Moreover, a higher concentration of sulphuric acid is required for extraction compared to hydrochloric acid. However, from nitric acid the extraction efficiency decreases as the concentration of the acid increases, due probably to oxidation of the reagent. But nitrate appears to be a preferred anion for the extraction of some elements.

Fluoride ion plays an important role, as does cyanide in suppressing the extractibility of some ions in preference to others.

Among the interesting features of the extraction of metal ions from hydrochloric acid with chloroformic BPHA is the wide variation of acid concentrations at which the different ions exhibit maximal extraction efficiency. While for chromium(VI), antimony(III) and tin(IV) the extractability gradually decreases with increasing acidity above 1 N, the extraction of titanium(IV), niobium, tantalum and vanadium(V) are found to increase with increasing acidity up to 11 N. Extraction maxima for technetium(VII) and tungsten(VI), on the other hand, are reached at only about half this acidity.

Some elements like zirconium and halfnium, which although are practically completely extracted from 1 N acid, like molybdenum, develop a minimum in the extraction curve at a certain higher acid region. This can be prevented by using a higher BPHA concentration. A wide acid range for the extraction of tungsten is also obtained with a higher reagent ratio.

From sulphuric acid alone the acidity required for maximal extraction of niobium, tantalum, vanadium(V), titanium(IV) and tungsten(VI) is again fairly high whereas that for zirconium and hafnium is quite low.

Another extractability difference is observed between antimony(III) and tin(IV). Whereas the former is extracted to the extent of 80% or 90% from 0·7 N to 34·0 N acid, the latter is 80% extracted only from 20 N acid. Again an increase in BPHA concentration raises its extractability to over 90% from 10 N acid.

A table depicting the optimal extraction pH values for BPHA complexes with various metal ions, as prepared by Shendrikar,[1] is given in Table 6.1.

The Separation of Thorium and Uranium from Lanthanum
(Dyrssen[2])

The distribution of thorium and uranium between chloroform and an aqueous 0·1 M perchloric acid–sodium perchlorate solution, and that of lanthanum between chloroform and an aqueous 0·1 M sodium

TABLE 6.1. OPTIMAL pH FOR EXTRACTION OF ELEMENTS BY BPHA

	1	2	3	4	5	6	7	8	9	10	11	12	13	14	15	16	17	18
	H																	He
	Li	Be(II) 6.5											B	C	N	O	F	Ne
	Na	Mg											Al(III) 3.6	Si	P	S	Cl	Ar
	K	Ca	Sc(III) 5.2	Ti(IV) 0–1	V(V) 6.5	Cr(III) 3	Mn(II) 10	Fe(II),(III) 5	Co(II) 10	Ni(II) 9	Cu(II) 3.6	Zn(II) 9	Ga(III) 3.1	Ge	As	Se	Br	Kr
	Rb	Sr	Y(III) 6	Zr(IV) 0–1	Nb(V) 3.5	Mo(VI) 1	Tc	Ru	Rh	Pd(II) 3	Ag(I) 1	Cd(II) 10	In(III) 5.3	Sn(II),(IV) 0–1	Sb(III),(V)* 0–1	Te	I	Xe
	Cs	Ba	La(III) 8	Hf(IV)*	Ta(V) 0	W(VI) 3	Re	Cs	Ir	Pt	Au	Hg(II) 8	Tl(I) 10.5	Pb(II) 9	Bi(III) 4	Po	At	Rn
	Fa	Ra	Ac															

Ce(III),(IV) 7, 0–1	Pr(III) 6	Nd(III) 6	Pm	Sm	Eu	Gd	Tb	Dy	Ho	Er	Tm	Yb	Lu
Th(IV) 4.5	Pa(V)*	U(VI) 3.5	Np	Pu(IV)*	Am	Cm	Bk	Cf	E	Fm	Mv	No	Lw

* Complete extraction takes place at high acid concentrations.

perchlorate solution at 25° as a function of the N-phenylbenzohydroxamate ion concentration in the aqueous phase, is such that the three elements can be extracted with a chloroform solution of BPHA. $1/N \log K$ values for lanthanum, thorium and uranium(VI) are $-4\cdot8$, $-0\cdot17$ and $-1\cdot57$, respectively. As lanthanum requires a much higher pH for appreciable extraction, uranium(VI) and thorium can be separated from lanthanum by a single extraction at pH 4·5 with a 0·1 M chloroformic BPHA solution. For a good separation of thorium and uranium, some degree of fractionation is required.

Procedure

Adjust the pH of the test soln. to 4·5 using 0·1 M perchloric acid and 0·1 M sodium hydroxide soln. and maintain the ionic strength of the 15-ml soln. at 0·1 M with sodium perchlorate. Add an 0·1 M BPHA soln. in alcohol-free, water-saturated chloroform. Shake to extract thorium and uranium, leaving lanthanum in the aqueous phase.

Simultaneous Separation of Iron and Small Amounts of Titanium from Aluminium
(Zharovskii[3])

Iron(III), titanium(IV), vanadium(V) and aluminium precipitate on the addition of an ethanolic BPHA solution in the pH ranges $-0\cdot6$–$10\cdot0$, $-0\cdot9$–$10\cdot0$, $-0\cdot8$–$7\cdot0$ and $3\cdot0$–$8\cdot0$, respectively. As these compounds can be extracted into chloroform, a method has been developed for the solvent extraction separation of iron, titanium and vanadium from aluminium. By repeated extraction from 0·5 N acid with a chloroformic BPHA solution, iron, titanium and vanadium are completely transferred to the organic phase, leaving only aluminium in the aqueous solution.

This method of separation can be applied effectively for the determination of aluminium in iron ores containing vanadium and titanium, after the dissolution of the ores in hydrochloric acid.

Procedure

Neutralize with ammonia the soln. containing iron (12–61 mg) and aluminium (2–35 mg), but only a small quantity of titanium (1–3 mg), in a separating funnel, until a turbidity appears. Dissolve the turbidity with a few drops of N sulphuric acid and add a further 5 ml of this acid. Dilute to 50 ml with water. Add a 5% ethanolic BPHA soln. (5 ml for every 10 mg of iron present) followed by chloroform (20 ml). Shake to extract the iron and titanium complexes into the chloroform layer. Repeat the extractions at least 2–3 times until a new portion of chloroform remains colourless. For the repeated extractions, add before each extraction with chloroform more BPHA soln. (1–2 ml). After all the iron and titanium, has been extracted, heat the aqueous phase on a water bath till the chloroform is completely removed. Determine the aluminium content of the soln. gravimetrically or volumetrically.

Separation of Scandium from the Lanthanides, Zirconium and Titanium
(Alimarin and Yun-hsiang[4])

Scandium-*N*-benzoylphenylhydroxylaminate is extracted into many organic solvents such as *n*-butanol, amyl acetate, benzene, chloroform and iso-amyl alcohol. An 0·5% solution of BPHA in iso-amyl alcohol completely extracts scandium at pH $\geqslant$ 4, whereas the extraction of the lanthanides begins at pH $>$ 5·8. The zirconium and titanium complexes with BPHA are extracted quantitatively from 1–8 N and 5–10 N hydrochloric acid, respectively, and can thus be separated from scandium at these high acidities. Sodium, ammonium, chloride and nitrate ions have no noticeable effect on the extraction of scandium.

Separation of Scandium from the Lanthanides and the Determination of Scandium

Transfer a chloride soln. of scandium and lanthanides (containing 0·025–0·075 mg of scandium oxide and 10–30 mg of lanthanides) to a 30-ml separating funnel. Adjust the pH to 4·5 with dilute hydrochloric acid or ammonia soln. Add an equal volume (2–5 ml) of an 0·5% BPHA soln. in iso-amyl alcohol and shake to extract all the scandium into the organic layer. Separate the organic layer from the aqueous phase.

To determine scandium, shake the organic layer twice with 5-ml portions of 1–2 N hydrochloric acid and transfer the acidic solns. to a 100-ml graduated flask. Add an 0·1% alizarin red S soln. (2 ml). Neutralize with (1+3) ammonia soln. until a red colour appears, and add a 20% ammonium acetate buffer soln. (pH 3·5, 5 ml). Cool to room temperature and dilute with water to make up to 100 ml. Measure the absorbance at 520 nm in a 5-cm cuvette.

Separation of Scandium from Zirconium or Titanium

Adjust the acidity to 2 or 5 N in hydrochloric acid and extract zirconium (3·2 mg of zirconium dioxide) or titanium (1 mg of titanium dioxide) respectively by shaking twice with 3% BPHA soln. in iso-amyl alcohol (5 ml). Scandium, remains in the aqueous phase.

Extraction of Aluminium from Complex Mixtures Including Uranium-based Fuels and Stainless Steels
(Villarreal, Krsul and Barker[5])

The method is based on the extraction of aluminium as a BPHA complex into benzene from an ammonium carbonate solution containing several masking agents. It may easily be back-washed into 0·2 N hydrochloric acid. Fluoride, EDTA and citrate interfere by masking aluminium and so must be absent.

Reagents

Thioglycollic acid. 80% aqueous soln.
Potassium cyanide. 5% aqueous soln.
BPHA. 2%, in ethanol or acetone.
Hydrogen peroxide. 1·2% aqueous soln.
Ammonium carbonate–hexametaphosphate buffer solution. Dissolve sodium phosphate $(NaPO_3)_6$ (5 g) and ammonium carbonate (200 g) in water (900 ml). Wash the soln. in a separating funnel 3 times with a 2% 8-quinolinol soln. in chloroform and twice with chloroform alone. Store the buffer soln. in a polyethylene bottle.

Procedure

Add to a 60-ml separating funnel an acidic soln. of the sample containing 5–30 μg of aluminium. Add to it, in succession, solns. of thioglycollic acid (1 ml), an excess (1–5 ml) of the ammonium carbonate–hexametaphosphate buffer over that required to make the soln. basic, potassium cyanide (1 ml) and hydrogen peroxide (1 ml) and dilute to 30 ml with water. Add the BPHA soln. (1 ml), mix and allow to stand for 5 min. Shake for 1 min with benzene (15–20 ml) to extract the aluminium–BPHA complex. Allow the phases to separate. Drain off the aqueous phase and wash the organic phase with water (10–15 ml). For the back-extraction, shake the benzene layer for 1 min with 0·2 N hydrochloric acid (15–20 ml).

Extraction of Thorium and Separation from the Lanthanides

(Alimarin and Yung-hsiang[6])

The thorium–BPHA complex can be extracted into chloroform, benzene, amyl acetate, n-amyl alcohol and iso-amyl alcohol. Thorium is completely extracted at pH 3 by a 3% solution of BPHA in iso-amyl alcohol. As the lanthanides are not extracted even at pH 5·8, thorium can be preferentially extracted at pH 4·5 leaving the lanthanides in the aqueous phase.

Procedure

To the thorium soln. (containing 9–26 mg of thorium dioxide) in a 30-ml separating funnel, add dilute ammonia soln. or dilute hydrochloric acid in suitable quantities to adjust the pH to 4·5 and the volume to 5 ml. Add a 3% BPHA soln. in iso-amyl alcohol (5 ml). Shake to preferentially extract thorium from the aqueous phase.

To determine thorium, transfer the organic phase to another separating funnel, extract thorium twice with 5-ml portions of 1–2 N hydrochloric acid and determine thorium spectrophotometrically with alizarin red S.

Extraction of Tungsten

(Che-ming, Chung-fen and Shu-chuan[7])

Microgram amounts of tungsten(VI) can be extracted quantitatively from 1–8 N hydrochloric acid or 14–22 N sulphuric acid with a chloroformic BPHA solution in as little as 2 min. Benzene, carbon tetrachloride, isobutyl methyl ketone and isopentyl alcohol can also be used as solvents. Up to 500 mg of uranyl nitrate, 10 mg of phosphate, 1 mg of iron, aluminium, fluoride or tartrate, 0·1 mg of chromium(VI), vanadium(IV), arsenic(III), antimony(III), bismuth or barium or 0·03 mg of lead, nickel, titanium or copper do not interfere. Molybdenum and vanadium are also extracted and zirconium or titanium (0·1 mg) and oxalate decrease the extent of extraction.

Procedure

To the tungsten(VI) soln. (containing 10 μg of tungsten), add hydrochloric or sulphuric acid to make the resulting 30-ml soln. 1–8 N in hydrochloric acid or 14–22 N in sulphuric acid. Add an 0·2% chloroformic

BPHA soln. (5 ml). Shake for 3–4 min. Tungsten is completely extracted into the chloroform layer.

Extraction of Zirconium

(Che-ming, Chung-fen and Shu-chuan[8])

The extractability of zirconium has been found to vary with changes in acid and BPHA concentration. From 0·3 to 10·0 N hydrochloric acid, zirconium is quantitatively extracted into a 0·6% solution of BPHA in chloroform. But with a weaker than 0·6% BPHA solution, as the acid concentration increases, the percentage of extraction decreases to a minimum at 4–5 N acid above which acidity it increases again.

From 10 ml of 1 N hydrochloric acid, 5–100 μg of zirconium is extracted quantitatively by 5 ml of an 0·1% BPHA solution in chloroform. But from 10 ml of 4 N hydrochloric acid, the extraction into 5 ml of an 0·2% BPHA solution in chloroform is complete on the addition of 0·1 g of ammonium thiocyanate or ammonium nitrate to the aqueous phase.

The finding that zirconium is quantitatively extracted by BPHA into chloroform from 0·3–10·0 N hydrochloric acid has been confirmed by studies on the extraction of trace quantities of hafnium and zirconium. The distribution ratios (log q) of hafnium and zirconium as a function of perchloric and hydrochloric acid concentrations for different concentrations of BPHA in benzene or chloroform show[9] marked minima on the extraction curves at 4–5 N acid. However, with increasing BPHA concentration ($2·5 \times 10^{-3}$ to $5·0 \times 10^{-3}$ M) the curves shift to higher q values. At certain BPHA concentration the q value at the minimum may even be higher than 100 and then the metal ions can be quantitatively extracted over the whole acidity range. At a still higher hydrochloric acid concentration (10–11 N), the decrease in q values is probably caused by the decomposition of BPHA.

From the extraction mechanism in strongly acidic solution (> 6 N), the extracted species have been suggested to be $MX_4 \cdot 2BPHA$ ($X = ClO_4^-$ or Cl^-). Later,[10,11] however, the complex extracted from perchloric acid (1·0–7·5 N) and hydrochloric acid (< 2 N) has been deduced to be $Hf(BPHA)_4$.

Separation of Niobium from Tantalum
(Alimarin, Petrukhin and Yun-hsiang[12])

When BPHA was used as a precipitant for the separation of niobium from tantalum when present in ratios of 1:100 to 100:1, at pH 3·5–6·5, it was observed[13] that the niobium precipitate was highly soluble in chloroform.

This observation was developed to give a highly satisfactory separation of niobium from tantalum by extraction of the niobium–BPHA complex into chloroform. From a tartrate solution, pH 4–6, $\geqslant 98\%$ of the niobium is extracted by a single extraction, leaving all the tantalum in the aqueous phase. At pH 0·5–3·0, tantalum is partially floated at the boundary of the phases and at pH 6–9 there is a slight but poorly reproducible extraction of tantalum. A successful separation of niobium from tantalum even when present in the ratios of 1:100 and 100:1 has been achieved by extraction from a sulphate solution containing 3% tartaric acid and maintained at pH 4·5–5·0.

Niobium and tantalum solutions. Heat a weighed quantity of mixed niobium and tantalum pentoxides in a mixture of sulphuric acid and ammonium sulphate. Dissolve the sulphates in tartaric acid soln. so that the final soln. is 3% in tartaric acid.

Procedure

Adjust the acidity of 4 ml of soln. (containing 0·1–10 mg of niobium pentoxide and 0·1–10 mg of tantalum pentoxide) to pH 4·5–5·0 in a small separating funnel. Add a 10% BPHA soln. in ethanol (1 ml) and chloroform (5 ml). Shake for 3 min and allow the layers to separate. The organic and the aqueous phases contain the niobium and tantalum, respectively.

The Separation of Niobium and Tantalum
(Alimarin and Petrukhin[14])

The extraction with chloroform of niobium and tantalum BPHA complexes as a function of acidity has been studied both from 3% tartaric acid solutions, from pH 12 to 20 N in sulphuric acid, and also from 3 to 30 N sulphuric acid. The niobium complex can be extracted from 12 to 24 N sulphuric acid. There is a minimum in the sulphuric acid–per cent extraction curve in the region 18–22 N acid if the shaking period is only 3 min. When this is increased to 15 min, the minimum disappears and over 99% of the niobium is extracted by a single extraction. From 20 N sulphuric acid about 80% of the tantalum is extracted. The investigation also confirms the previous findings[12] that at pH 4–6 while niobium is extracted, tantalum remains in the aqueous phase and at pH 6–9 the extractability of the tantalum complex is poor and irreproducible.

The extraction of both niobium and tantalum at pH 1–12 is dependent on the method of introducing BPHA. With a chloroform solution of BPHA the extraction is less than when ethanolic solution of BPHA is first added before extraction with chloroform. With the ageing of the tantalum solution, tantalum extraction at pH 2·4 is reduced.

Procedure

Follow the procedure as described on p. 166 for extractions from 3% tartaric acid soln. ranging from pH 12 to 20 in sulphuric acid and containing 0·35 mg of niobium pentoxide or 0·30 mg of tantalum pentoxide per ml, using a 5% ethanolic BPHA soln. (1 ml) and a volume of chloroform equal to the total volume of the soln.

For extraction from solns. containing no tartaric acid, dilute the 32 N sulphuric acid (containing 2·8 mg of niobium or tantalum pentoxide per ml) to desired volumes to prepare solns. of different concentrations. Place any of the solns. or suitable aliquots in a glass-stoppered tube and shake to extract with an equal volume of a 1% soln. of BPHA in chloroform.

Extraction of Niobium, Tantalum, Titanium, Zirconium and Vanadium from Sulphuric Acid
(Alimarin and Petrukhin[15])

The elements that easily hydrolyse at pH > 0 often form polymeric and colloidal particles in such solutions. In such instances, as with niobium and tantalum, where addition of complexing agents does not always eliminate the formation of colloids and polymers, extractions of such elements as complexes are more effectively achieved from strongly acidic solutions. Complex formation at low acidity is often not very selective. Thus a study on the extraction from purely sulphuric acid solutions of niobium, tantalum, titanium, zirconium and vanadium with a 0·1 M BPHA solution in chloroform as well as of BPHA itself has been made.

The distribution curve shows that BPHA is 100% extracted from up to 18 N sulphuric acid and thereafter, as the acidity is increased, the extraction decreases. At > 26 N the reagent is no longer extracted into chloroform.

Niobium (2.7×10^{-3} M) begins to be extracted from a 3 N sulphuric acid solution and with the increase in the concentration of the acid the percentage of niobium extraction increases till at 12–24 N acid 99·9% of the niobium is extracted after shaking for 15 min. About 80% of the tantalum (1.4×10^{-3} M) is extracted from 20 N sulphuric acid but an increase or decrease in the acid concentration reduces the per cent extraction.

Zirconium (1.1×10^{-3} M) extraction is complete at an acidity of 1–2 N; as the acidity increases to 6–7 N the extraction decreases to 70% and then is constant until the acid concentration is 24–25 N. With the further increase in the acid concentration, there is a sharp decrease in extraction. Titanium (2.1×10^{-3} M) is completely extracted from 12–22 N sulphuric acid. But when the acid concentration is > 22 N, the percentage extraction decreases markedly. From 18–22 N sulphuric acid, vanadium extraction is complete. When, however, the concentration of the acid is decreased to 14 N, the extraction decreases to only 8%. It increases again as the acid concentration is lowered further. The normalities of sulphuric acid required for 50% extraction of BPHA and its metal complexes are: 22 N for BPHA, 27 N for niobium and zirconium, 25 N for vanadium and 23·2 N for titanium.

In sulphuric acid the BPHA complexes are intensely coloured. From a study of the composition (at 400–440 nm) of the niobium complex in 30 N sulphuric acid by a mole ratio method it appears that BPHA and niobium combine in a ratio of 3:1, which corresponds to the formula $NbO(C_{13}H_{10}O_2N)_3$ suggested previously.[13] The solubility of the BPHA complexes in 30 N sulphuric acid suggests that in the complexes, the BPHA nitrogen acts as a weak base to combine with a hydrogen ion so that the complex exists as a cation.

Separation of Protactinium from Niobium, Titanium, Zirconium and Hafnium

(Palshin, Myasoedov and Novikov[16])

The BPHA complex of protactinium-233 is quantitatively extracted by benzene from hydrochloric and sulphuric acids over a wide range of acid concentration. From nitric acid, however, the amount of protactinium extracted decreases with increasing acid concentration. By extraction from 4–18 N sulphuric acid with a 0·1 M BPHA solution in benzene, greater selectivity of protactinium extraction is achieved, since very few elements are extracted under these conditions. Niobium, titanium, zirconium and hafnium are not extracted if complexing agents like hydrofluoric acid, oxalic acid and hydrogen peroxide are used to keep these elements in the aqueous phase.

Separation of Protactinium from Other Elements

(Myasoedov, Palshin and Palei[17])

From a study of the extraction of protactinium from sulphuric and hydrochloric acid media with a 0·1 M solution of BPHA in benzene, it has been observed that the presence of up to 15 mg of chromium, aluminium, iron, bismuth, manganese, nickel, tin(II and IV), titanium(III), lanthanum, cerium(IV), thorium or uranium per ml does not interfere with the extraction of protactinium from 7 N sulphuric acid. The yellow protactinium(V)–BPHA solution in benzene shows an absorption maximum at 365 nm and obeys the Lambert–Beer law over the range 1–6·9 μg of protactinium per ml of benzene. The molar absorptivity is 8400. Isolation of protactinium is given below.

Procedure

Dissolve neutron irradiated thorium nitrate (100–200 mg) in water (5 ml) and 7 N sulphuric acid (8 ml), add a 0·1 M BPHA soln. in benzene (10 ml), shake for 1 min, set aside for 2 min and then separate the organic phase from the aqueous phase. Reject the aqueous phase. Wash the organic phase twice with an equal volume of a soln. which is 7 N in sulphuric acid and 0·1 M in oxalic acid. Discard the wash solns. Add to the washed extract a soln. which is 7 N in sulphuric acid and 0·06 M in hydrofluoric acid (10 ml) and back-extract the protactinium.

Separation of Niobium from Zirconium
(Lyle and Shendrikar[18])

Niobium-95, in a carrier-free form, has been separated from zirconium by extraction with a chloroformic solution of BPHA from hydrochloric acid containing a little fluoride, which prevents the extraction of zirconium. The percentage of niobium extracted is dependent on hydrofluoric acid concentration and is also a function of time. For an efficient extraction in a reasonable time, the minimum concentration of BPHA required is 0·2% in chloroform. The extraction method given below gives a decontamination factor of about 10^4 for niobium from zirconium. The presence of up to 1 mg per ml of zirconium does not affect the separation. From the chloroform phase, niobium can be back extracted into *aqua regia*, aqueous ammonia or ammoniacal hydrogen peroxide. The entire procedure does not take more than 30 min.

Extraction of Niobium into Chloroform

To the soln. of zirconium and niobium (which may contain up to 1 mg of zirconium per ml) add enough hydrofluoric and hydrochloric acids to make the soln. 0·05 M in hydrofluoric acid and 1 M in hydrochloric acid. To this soln. in a separating funnel, add an equal volume of 0·2% BPHA soln. in chloroform and shake for 12 min. Allow the two phases to separate, and, if necessary, repeat the extraction of niobium with the chloroformic BPHA solution. Combine the extracts. Wash the chloroform extract twice with an 0·05 M hydrofluoric acid–1 M hydrochloric acid mixture. Discard the washings.

Back Extraction of Niobium

Any of the following procedures can be used:

(i) With *Aqua Regia*

Add an equal volume of *aqua regia* to the chloroform extract, shake for 10 min, remove the aqueous phase and wash it twice with 10-ml portions of chloroform. Evaporate it to 5–6 ml, cool, extract twice with 2 small portions of diisopropyl ether to remove any residual organic matter and warm to drive off the ether.

(ii) With Ammoniacal Hydrogen Peroxide

Shake the chloroform phase for 10 min with an equal volume of 7 M hydrogen peroxide soln., adjusted to pH 11 with dilute ammonia soln.

Discard the chloroform phase, remove the aqueous phase and treat it as in (i) up to the removal of organic matter. Add *aqua regia* to oxidize ammonia and finally prepare the soln. in hydrochloric or nitric acid.

(iii) With Ammonia

Shake the chloroform extract for 10 min with 6 M ammonia soln. Remove the aqueous phase and treat it as described in (i) to remove organic matter and much of the ammonia. Oxidize the residual ammonia with *aqua regia* and prepare a soln. in hydrochloric or nitric acid.

Note

The separation of niobium from zirconium can also be undertaken in the presence of sulphuric, tartaric or oxalic acids. However, in the presence of oxalate, the hydrochloric acid concentration must be increased, preferably to 6–7 M for a M oxalate soln.

Separation of Tin and Antimony from Indium
(Rakovskii and Petrukhin[19])

Extraction studies have been made with the use of radioactive isotopes of ^{114}In, ^{122}Sb and $^{113-125}$Sn and by measuring the activities of the organic and the aqueous phases obtained on extraction with an equal volume of a 0·05 M chloroformic BPHA solution from a 5 ml solution having the requisite sulphuric or hydrochloric acid concentration. From hydrochloric acid the percentage extraction of tin(IV) and antimony(III) decreases with increasing acid concentration. The antimony(III) extraction curve passes through a minimum at 5·0–8·5 N acid and on further increase of the acidity, the percentage extraction increases to only 20% at 11 N acid. But from sulphuric acid solns. of conc. 0·7–34·0 N 80–90% of the antimony(III) is extracted; with over 20 N sulphuric acid, antimony(III) extraction takes place even in the absence of BPHA.

For tin(IV), 80% extraction is achieved from 20 N sulphuric acid solution. If, however, the concentration of BPHA is increased to 0·4 M, the percentage of tin extraction is more than 90% even from 10 N acid.

As indium remains almost completely in the aqueous phase when the sulphuric acid conc. is higher than 0·4 N, an attempt to separate tin(IV) and antimony(III) from indium(III) has been made from sulphuric acid of higher concentration.

Procedure

Dissolve the sample in a mixture of conc. sulphuric acid and ammonium sulphate. Adjust the volume to 30 ml keeping its sulphuric acid conc. at 10–12 N. Add a 0·4 M soln. of BPHA in chloroform (10 ml) and shake to extract. Repeat the extraction operation.

Separation of Gallium, Indium, Thallium, Germanium, Tin and Lead
(Lyle and Shendrikar[20])

In the absence of BPHA, germanium can be extracted into chloroform from $\geqslant$ 8 M hydrochloric acid. With a 1% BPHA solution in chloroform, tin(IV) is efficiently extracted from 0·8 M hydrochloric acid or 4 M perchloric acid whereas gallium, indium, thallium(I) and lead are extracted from acetate-buffered solutions. From the pH–per cent extraction curve, it appears that gallium can be separated with maximal efficiency from indium, thallium(I) and lead by extraction at pH 3·1–3·8, indium from lead and thallium(I) at pH 5·3–5·6 and lead from thallium(I) at pH 7·8–8·2. Thallium(I) extraction is quantitative above pH 10·5, though the extraction begins at pH > 8·2. The results, recorded in Table 6.2, show that all six elements can be separated from each other by a proper choice of conditions. The per cent extraction given in the table refers to a single extraction with equal volumes of chloroformic and aqueous phases.

TABLE 6.2. OPTIMAL CONDITIONS FOR THE EXTRACTION OF CERTAIN METAL IONS WITH BPHA

Ion	pH	% BPHA in chloroform	Equilibration time (min)	Maximal metal ion conc. (mg/10 ml)	% Extraction
Gallium	3·1	1·0	12	10	99·4
Indium	5·3	1·0	10	10	99·2
Thallium(I)	10·5	0·7	8	8	99·7
Germanium	8·0 M HCl	0·0	15	10	99·8
Tin(IV)	0·8 M HCl	1·0	10	10	94·4
	4·0 M HClO$_4$	1·0	10	10	96·1
Lead	9·0	1·0	10	12	97·4

Separation factors for pairs of elements are: (i) 10^4 for gallium from indium, indium from lead, gallium from lead and tin from lead; (ii) 10^3 for gallium from germanium, indium from tin, tin from gallium and tin from indium; and (iii) 6×10^2 for lead from thallium.

Reagents

Chloroform, containing 2% of ethanol as a preservative.

Acetate buffer solution, pH 5·3. Mix anhydrous sodium acetate (80 g) with glacial acetic acid (8 ml) and dilute to 1 litre.

Procedure

To the gallium, indium, thallium(I) or lead soln. add the buffer soln (0·5 ml) and adjust the pH to the required value (see text) by the addition of hydrochloric acid or ammonia, so that the final volume is 10 ml. For tin(IV), make the 10-ml soln. 0·8 N hydrochloric acid or 4 M in perchloric acid. Add an equal volume of a 1% soln. of BPHA in chloroform, equilibrate for 15 min and allow the phases to separate. For germanium, adjust the acidity to 8 M in hydrochloric acid and shake for 15 min with an equal volume of chloroform alone to extract the metal ion.

During the separation of pairs of elements, follow the procedure as suggested above but extract the aqueous phase twice with equal volumes of chloroform phase and then wash the aqueous phase twice with equal volumes of chloroform. Also wash the organic phase twice with aqueous solns. of appropriate pH but without the metal ions.

Notes

1. By simple adjustment to a much lower pH, gallium, indium, thallium(I) and lead can be back extracted into the aqueous phase. Tin(IV) and germanium are back extracted, respectively, into 3M hydrochloric acid saturated with oxalic acid and into < 4 M hydrochloric acid or dilute perchloric acid.

2. Iron(III)- and aluminium-BPHA complexes are soluble in chloroform, and will interfere in the separations. Oxalic and tartaric acids must be absent, as they also interfere.

Separation of Protactinium from Niobium, Tantalum and Other Elements

(Lyle and Shendrikar[21])

Investigations of the separation by solvent extraction of protactinium from niobium, tantalum and other elements show that the efficiency of extraction of protactinium, niobium and tantalum into chloroform depends on the concentrations of fluoride, hydrochloric acid and BPHA present. For maximal extraction, protactinium requires a 1% BPHA solution, 0·025 M fluoride and > 10·5 M hydrochloric acid. Niobium requires an 0·2% BPHA solution when the fluoride concentration is 0·05 M and the hydrochloric acid is of any molarity. For tantalum, in 0·4 M fluoride and ⩾ M hydrochloric acid, a 0·5% BPHA solution is required. Tantalum extraction, unlike that of niobium, is dependent on hydrochloric acid concentration below 1 M acid. The times required for maximal extraction under these conditions are 12, 9 and 6 min, respectively, for niobium, tantalum and protactinium. From radio-tracer quantities to 1 mg of tantalum per ml, 0·5 mg of niobium per ml and up to a few tenths of a mg of protactinium per ml are efficiently extracted.

Thus from 1 M hydrochloric acid 0·4 M in fluoride, tantalum can be separated from protactinium. Similarly, niobium can be separated from protactinium from a solution 1 M in hydrochloric acid and 0·05 M in fluoride and from tantalum from a solution 0·01 M in hydrochloric acid and 0·05 M in fluoride. As aluminium, thorium and the lanthanides are not extracted at high acidity (10–11 M), protactinium may be separated from them at such acidities.

Niobium can quantitatively be recovered from the chloroform layer by a single back extraction with *aqua regia*, ammonia (6 M) or hydrogen peroxide solution (7 M) whose pH has been adjusted to 10–11 by dilute ammonia solution. Tantalum (99·7%) can also be back extracted with ammonia solution (2·5 M) and also (99·8%) with hydrogen peroxide solution (2 M) at pH 10–11. Quantitative recovery of protactinium can be made by a single back extraction with oxalic acid solution (M), hydrogen peroxide solution (7 M) at pH 10–11 or with hydrofluoric acid (0·6 M), but only 95·7% is recovered by a tartaric acid solution (2 M). The optimal time required for the back extraction depends both on the metal ion and the chemicals used. For instance, for back extraction of niobium, *aqua regia* and hydrogen peroxide need 15 min, whereas ammonia requires 12 min. For tantalum, hydrogen peroxide needs 12 min, but ammonia takes only 8 min. For protactinium, oxalic and hydrofluoric acids require 9 min, tartaric needs 6 min and hydrogen peroxide takes 12 min.

By comparison of the γ-ray spectra of pairs of isotopes and of single isotopes with those obtained from fractions after separation from mixtures of the metals, the separation factors are found to be 10^3 for the systems tantalum (1 mg/ml)–protactinium, niobium (0·5 mg/ml)–protactinium and for niobium (0·5 mg/ml)–tantalum. A separation factor of 10^2 is obtained, if niobium and tantalum are both extracted, and then niobium (0·5 mg/ml) is back extracted with *aqua regia* to separate it from tantalum.

Separation of Tantalum from Protactinium

Add hydrochloric acid and potassium fluoride so that on dilution the 10-ml soln. containing tantalum and protactinium is 1 M in hydrochloric acid and 0·4 M in fluoride. Equilibrate in a polyethylene bottle for 9 min with an equal volume of a 0·5% BPHA in chloroform. Separate the chloroform layer from the aqueous phase in a separating funnel. Repeat the extraction procedure with the same volume of the chloroformic BPHA soln. Collect the aqueous and chloroform phases separately and wash each of the phases twice, the chloroform phase with 10-ml portions of a solution which is 1 M in hydrochloric acid and 0·2 M in fluoride and the aqueous phase with 10-ml portions of chloroform. Mix the washings with the appropriate fractions. The chloroform phase contains all the tantalum while the aqueous phase contains the protactinium.

Separation of Niobium from Protactinium

Follow the method given above with 10 ml of soln. containing niobium and protactinium which is 1 M in hydrochloric acid and 0·05 M in fluoride. Use 0·2% BPHA in chloroform for extraction, with an equilibration time of 12 min.

Separation of Niobium, Tantalum and Protactinium

Adjust the acidity and the fluoride concentation of the solution containing niobium, tantalum and protactinium such that the 10-ml of soln. is 1 M in hydrochloric acid and 0·05 M in fluoride. Add 10 ml of an 0·5% BPHA soln. in chloroform. Shake for 12 min to extract niobium and tantalum. Separate the layers and repeat the extraction procedure with the aqueous phase. The chloroform layer now contains all the niobium and tantalum and the aqueous layer contains the protactinium.

For the subsequent separation of niobium from tantalum, shake the chloroform extract for 15 min with *aqua regia* to remove niobium into the aqueous phase.

For a more efficient separation of niobium, back extract both niobium and tantalum by shaking with a 6 M ammonia soln. Adjust the hydrochloric acid and the fluoride concentrations of the soln. to 0·01 M and 0·05 M, respectively, and extract just the niobium into the chloroform layer by shaking for 12 min with 0·2% BPHA in chloroform.

Separation of Niobium and Tantalum Fluorocomplexes
(Erskine, Sink and Varga[22])

By a radiotracer technique, the extraction behaviour of trace and millimolar concentrations of tantalum and niobium as fluorocomplexes with BPHA in chloroform has been studied as a function of fluoride and perchloric acid concentrations. From the distribution data, the extracted species are calculated to be $TaF_3 \cdot 2BPA$ and $H_2NbF_5 \cdot 2BPA$ with extraction equilibrium constants of 2×10^{10} and 3×10^7, respectively, using 10^{-3} M BPHA solution.

As the fluoride ion concentration is increased to 10^{-4} M, the extraction of both metals decreases; the niobium extraction decreases more than that of tantalum. Moreover, tantalum extraction is dependent on perchloric acid concentration. But this dependence decreases as the acidity increases, with the smallest difference occurring between 4 M and 5 M acid. However, such a systematic dependence on acidity is not observed with the niobium extraction.

In the absence of perchlorate, single stage decontamination factors ranging from 51 to 70 at fluoride concentrations of $2 \cdot 9 \times 10^{-5}$ to $1 \cdot 0 \times 10^{-3}$ M are obtained when the elements in millimolar concentrations are extracted into a $2 \cdot 5 \times 10^{-2}$ M BPHA solution in chloroform.

From these findings it appears that with BPHA as a chelating ligand, efficient separation of niobium and tantalum can be achieved by multiple or continuous solvent extraction.

Separation of Plutonium from Uranium, Americium, Zirconium and Other Fission Products
(Chmutova, Petrukhin and Zolotov[23])

Plutonium(IV) is quickly extracted as its BPHA complex into chloroform from 1–6 M nitric acid after only brief contact between the organic and aqueous phases. Because BPHA decomposes in

4–6 M nitric acid, it is preferable to extract from 3 M nitric acid, under which condition a 0·4 M BPHA solution in chloroform readily gives 100% extraction of plutonium when present in the range 0·002 to 1·0 mg/ml. The presence of acetate does not affect the extraction but an increase in the sulphate or oxalate concentration decreases considerably the degree of plutonium extraction.

Though plutonium(IV) is extracted noticeably from hydrochloric acid by a 0·4 M chloroformic BPHA solution after a short time, it remains in the aqueous phase if the contact time is increased to 1 hr. Plutonium thus once extracted into the organic phase from nitric acid can be completely back extracted from the organic phase by 5 M hydrochloric acid after a contact time of 1 hr. This is due to the reduction of plutonium(IV) to plutonium(III) by hydrochloric acid.

A similar attempt to extract plutonium from sulphuric acid by a 0·4 M BPHA solution in chloroform was not successful. Likewise, when a 5-ml portion of the plutonium extract in chloroform obtained from a 3 M nitric acid solution is shaken with 3·5 M sulphuric acid, with the volume of the organic and water phases in a ratio of 2·5:1, plutonium is completely back extracted into the sulphuric acid.

Under the optimal conditions for plutonium(IV) extraction with 0·4 M chloroformic BPHA solution, uranium(VI), americium and neptunium(V) are practically unextracted, like the other disintegration products such as ruthenium-106, lanthanum-140 and cerium(IV)-144. Zirconium-95 and niobium-95, however, are appreciably extracted.

Separation of Plutonium(IV) from Uranium(VI) and Americium(III)

To a 5-ml 3 M nitric acid soln. of plutonium ($8·8 \times 10^{-6}$ M) and uranium (^{233}U + ^{238}U, 10–1000 times the amount of plutonium) in a stoppered burette add a 0·4 M BPHA soln. in chloroform (5 ml), shake for 30 sec to extract plutonium into the chloroform and separate the organic phase from the aqueous phase. Plutonium, in the organic phase, can be determined by the α-emission from an aliquot of the organic phase. Uranium and americium remain in the aqueous phase.

Separation of Plutonium from Fission Products

Dissolve the neutron-irradiated uranium dioxide, aged for 1 year, in 4 M nitric acid. Dilute an aliquot in a graduated flask to make the soln. 3 M in acid. Repeat the procedure for separating plutonium described above. Zirconium and niobium are also extracted into the chloroform.

Plutonium can be separated from them by back extraction of the 5-ml extract by adding 3·5 M sulphuric acid (2 ml). Shake for 3 min, separate the aqueous phase, and wash it with pure chloroform. Plutonium, all of which is in the aqueous phase, can be determined radiometrically.

Separation of Protactinium from Neutron-Irradiated Thorium
(Lyle and Shendrikar[24])

Protactinium-233 ($t_{\frac{1}{2}}$ = 27·4 days) produced by slow neutron irradiation of thorium-232 can be extracted[21] into a chloroformic BPHA solution from strong hydrochloric acid not more than 0·025 M in fluoride. Thorium and uranium, however, are not significantly extracted[2] into a 0·1 M BPHA solution in chloroform from aqueous solutions of pH < 0·5 and 1·5, respectively. Thus protactinium is easily separated from them. Thorium daughters are either not extracted at all (radium-228, radium-224) or begin to extract at a pH of about 4.5 or higher (actinium-228 and lead-212) and hence do not extract with the protactinium. Alkali and alkaline earth metals, lanthanides, silver, cadmium, indium, ruthenium(III), rhodium(III), palladium(II), tellurium(IV) and (VI) are not expected to extract.

The separation factors for the separations of protactinium from thorium, uranium, tin and antimony are 3×10^4, $> 4 \times 10^3$, 7×10^2 and 2×10^2, respectively.

Separation of Protactinium-233 from Thorium and Uranium

Dissolve the irradiated thorium (metal, basic carbonate or oxide) in conc. hydrochloric acid containing a little fluoride. Heat, if necessary. Adjust the acidity and the fluoride concentrations to be ⩾ 10 M and 0·025 M, respectively. Add an equal volume of a 1% soln. of BPHA in chloroform and shake for 6 min. Separate the phases and shake the aqueous phase again with an equal volume of the chloroformic BPHA soln. Discard the aqueous phase. Combine the chloroform phases, and wash twice with equal volumes of conc. hydrochloric acid. Discard the washings. Shake the chloroform phase with an equal volume of a soln. which is 5 M in hydrochloric acid and ≮ 0·6 M in hydrofluoric acid. Wash the aqueous phase, which now contains the protactinium, with some chloroform.

If tin(IV) and antimony(III) are present, add to the final aqueous phase after back extraction, boric acid or aluminium chloride to remove the fluoride. Reduce the acidity to 0·8–1·0 M and extract these metal ions into a 1% BPHA soln. in chloroform. Wash the aqueous phase containing protactinium with some chloroform. If the presence of boron or aluminium in the aqueous phase is not desirable, make the aqueous phase strongly acidic in hydrochloric acid and repeat the procedures as described for the extraction of protactinium into the chloroform phase and its back extraction into the aqueous phase.

Separation of Arsenic, Antimony, Bismuth and Tin
(Lyle and Shendrikar[25])

The optimal conditions for the extraction of the elements are given in Table 6.3.

TABLE 6.3. OPTIMAL CONDITIONS FOR THE EXTRACTION OF ARSENIC, ANTIMONY, BISMUTH and TIN WITH BPHA

Ion	Acidity	% BPHA in chloroform	Equilibration time (min)	Maximal metal ion conc. (mg/10 ml)	% Extraction†
Tin(IV)	0·8 M HCl	1·0	10	10	94·4
	4·0 M HClO$_4$	1·0	10	10	96·1
Antimony(III)	1·0 M HCl	0·5	10	10	93·1
Antimony(V)	9·4 M HClO$_4$	0·9	15	8	98·3
Bismuth(III)	0·1 M HCl	0·6	10	8	99·4
Arsenic(III)	11·4 M HCl	0·0	7	12·5	91·2

† For a single extraction with equal volumes of aqueous and chloroformic phases.

A number of separations are possible. Tin(IV) can be separated from antimony(V) by extraction of the former from 0·8 M hydrochloric acid and arsenic(III) can be separated from bismuth(III) by partition with chloroform alone from strong hydrochloric acid. To separate bismuth from tin(IV), extract the latter from 4 M perchloric acid with BPHA. For the separation of antimony(V) from bismuth, extract antimony from ⩾ 9 M perchloric acid. Bismuth can be separated from tin(IV) and antimony(V) by extraction of the last two elements from a solution 9 M in perchloric acid and 0·1 M in oxalic acid. For each separation, extract the aqueous phase twice with equal volumes of the chloroformic BPHA and twice with chloroform alone.

The separation factors for tin(IV) from antimony(V), tin(IV) from bismuth(III) and antimony(V) from bismuth(III) are 1×10^2, 9×10^2 and 7×10^2, respectively.

Extraction of Thallium as a Function of pH and BPHA Concentration
(Schweitzer and Norton[26])

From 10 ml aqueous perchlorate solutions, 10^{-6} M in thallium(I)-204 and 0·1 M in sodium perchlorate, thallium(I) has been extracted at pH values ranging from 1 to 13, with equal volumes of 10^{-1} M and 10^{-2} M BPHA solutions in chloroform after equilibration for 3 hr at $30.0 \pm 0.5°$. On the basis of the results obtained, it was concluded that the extracting species is TlR (RH represents BPHA) and the aqueous species, as the pH rises, are Tl^+, TlR and TlR_2^- and that the pH values at which log D (D is the distribution coefficient) equals zero are 8·3 and 9·3 for 10^{-1} M and 10^{-2} M BPHA solutions, respectively. The log D values found at various pH values are given in Table 6.4.

Separation of Bismuth from Lead
(Chwastowska and Braginska[27])

As little as 5–10 μg of bismuth can be separated from lead by extraction with a solution of BPHA in chloroform. The methods for the separation and determination of bismuth in copper–silver alloy and in lead are given below.

Procedure

Dissolve the alloy (1 g) in (1 + 1) nitric acid (5 ml). Dilute to 50 ml in a graduated flask. Take an aliquot containing 10 μg of bismuth, adjust its acidity to pH 2 and extract the bismuth by shaking for 3 min with 5-ml, 2·5-ml and 2·5-ml portions of an 0·01 M BPHA soln. in chloroform. Combine the extracts. Back-extract bismuth from the organic phase by shaking for 5 min with 5 N nitric acid (5 ml). Determine bismuth in the aqueous phase spectrophotometrically with dithizone.

TABLE 6.4. VARIATION OF LOG D WITH pH FOR EXTRACTION
OF THALLIUM(I) WITH BPHA

BPHA, 0·1 M		BPHA, 0·01 M	
pH	log D	pH	log D
5·8	−2·47	5·5	−3·66
6·7	−1·64	7·0	−2·21
7·1	−1·18	8·2	−1·13
7·6	−0·70	9·2	−0·09
8·3	−0·01	9·5	0·12
8·4	0·16	9·8	0·49
8·8	0·65	10·6	1·17
9·3	0·99	10·8	1·21
9·8	1·18	11·0	1·20
10·1	1·22	11·8	1·20
10·4	1·21	12·0	1·19
10·8	1·19	12·1	1·21
11·2	1·21		
11·6	1·20		
11·9	1·00		
12·2	0·78		

If the lead content is more to give colour to dithizone, adjust the pH of the final aqueous soln. to 2 and repeat the entire extraction procedure.

For the determination of 5–10 μg of bismuth in lead samples, dissolve the sample (1 g) in 7 N nitric acid (20–30 ml). Dilute to 100 ml, warm and precipitate lead as its sulphate with 3·6 N sulphuric acid. Allow to settle for 3 hr. Decant the soln. Mix the precipitate with 5 N nitric acid (10 ml) and heat for 1 hr. Cool, filter, and wash the precipitate several times with (1 + 100) nitric acid. Evaporate the combined filtrates to 50 ml, filter off any precipitate of lead sulphate, dilute the filtrate in a graduated flask to 100 ml, take an aliquot containing 5–10 μg of bismuth and extract as above.

Extraction of Copper, Iron, Lead, Nickel, Cobalt and Cadmium

(Chwastowska and Minczewski[28])

At a 1:50 mole ratio of metal to BPHA, copper, lead, iron, nickel and cobalt are extracted almost quantitatively into chloroform under different pH conditions. The pH–per cent extraction curves show that iron and copper are extracted from slightly acidic solutions, whereas lead, nickel, cobalt and cadmium are extracted from alkaline solutions. The extraction curve for iron shows two maxima at pH 3 and 6 and a minimum at pH 4. Even on prolonged extraction cadmium is extracted only to the extent of 78%.

Procedure

Dilute a 0·001 M soln. (1 ml) of a metal ion to 5 ml. Adjust the pH to 3–11 by adding an acetate buffer (1 ml), prepared by mixing, as required, 1 N acetic acid with ammonia soln. Add a 0·01 M chloroformic BPHA soln. (5 ml). Shake for 60 sec. Separate the chloroform phase. Repeat the extraction procedure twice with 2·5 ml of the chloroformic BPHA soln., shaking each time for 30 sec. Combine the chloroform extracts which contain the metal ion.

Extraction of Zinc, Mercury, Bismuth, Manganese, Aluminium, Chromium and Silver

(Chwastowska and Minczewski[29])

Using a 50-fold excess of BPHA and by extractions from an aqueous solution with a chloroformic BPHA solution, with aqueous to organic volume ratios of 1:1 and 2:1, as in the previous experiment,[28] the above metals are extracted under different pH conditions. The extraction is repeated twice or thrice except for zinc and bismuth which are quantitatively extracted by a single extraction for 1 min, but aluminium takes 5 min for its complete extraction by a double extraction. Lead, nickel, cobalt, cadmium, zinc, mercury and manganese are extracted from alkaline solutions only and copper, iron and bismuth are extracted from acidic solutions. The optimal pH for quantitative extraction is 9 for zinc, 4 for bismuth, 10 for manganese and 6 for aluminium. The maximal extraction of mercury is only about 88% at pH 8·5, that of silver is about 33–39% at pH $\leqslant 1$ and that of chromium is 24% at pH 3–4. The extraction curve of manganese exhibits maxima at pH 2 and 6 and minima at pH 4 and 7. An acetate–ammonia buffer interferes with the recovery of mercury, as does chloride with that of silver.

Extraction of Several Common Metal Ions
(Chwastowska[30,31])

With an 0·01 M BPHA solution in chloroform, aluminium and iron are extracted from an aqueous nitrate medium at pH < 7, lead, nickel, cobalt, zinc and manganese are extracted at pH > 7 and copper and bismuth at the pH ≲ 7. Nitrate is the preferred anion as chloride and sulphate depress the extraction of cobalt, manganese, lead and nickel. In the presence of cyanide only lead, bismuth and aluminium are extracted and in presence of citrate only copper is extracted quantitatively. EDTA must be absent as in its presence no extraction is possible. The metals are determined spectrographically either on evaporation of the chloroform extract on graphite powder followed by excitation in the crater of a graphite electrode or by the graphite spark method after back extraction into nitric acid.

The optimal pH and maximum extraction percentages of thirteen metallic elements are as follows:

Cu 4–11, 100%; Fe 2–3, 100%; Pb 7–11, 97%; Ni 9–11, 97%; Co 10–11, 100%; Cd 10–11, 78%; Zn 9–11, 100%; Hg 7·5–9·0, 88%; Bi 2–13, 100%; Mn 10, 100%; Al 6, 100%; Cr 3–4, 23%; Ag 1, 35%. The effect of anions on some elements have been studied at pH 3–11 but for cyanide at 7–11.

Procedure

Take the test soln. (2 ml containing 10 μg of the metal ion per ml). Adjust the pH to the desired value by adding a 1 N acetate–ammonia buffer soln. (2 ml). Shake to extract the metal ion for 10 min with an equal volume of 0·01 M BPHA soln. in chloroform. Concentration on anion is 2% for EDTA, 10% for others and in the absence of anion nitrate is the only medium.

Extraction and Determination of Beryllium in Silicates
(Mareva, Puzdrenkova and Stoyanova[32])

Beryllium is extracted as its BPHA complex at pH ⩾ 8 to the extent of about 90% into isobutyl alcohol and to about 99% in *n*-butyl alcohol or ethyl acetate. It may be determined in the presence of aluminium, magnesium, manganese, calcium and iron.

Procedure

Heat the sample (0·10–0·15 g) with hydrofluoric acid in a platinum crucible. Evaporate to fumes 3 times with (1 + 1) sulphuric acid. Moisten the residue with a few drops of sulphuric acid, dissolve in 5% sulphuric acid on a water bath and dilute to 100 ml. Treat a 5-ml aliquot for 1–2 min at pH 4–5 with 0·05 M 1,2-cyclohexanedinitrilotetraacetic acid soln. (3 ml), buffer to pH 9 with a borax soln. (1 ml), add a 5% ethanolic BPHA soln. (1 ml) and extract 3 times with n-butanol. Combine the extracts and wash out beryllium from the combined extracts with 0·1 N hydrochloric acid and determine beryllium photometrically with Beryllon II in the hydrochloric acid soln.

Separation of Elements from Hydrochloric Acid. Application to the Extraction of Niobium and Zirconium from Uranium

(Vita, Levier and Litteral[33])

A solution of BPHA in chloroform has been used to study the extraction characteristics of the elements from 1–11 N hydrochloric acid by the procedure given below. The elements whose extraction efficiencies are greater than 50% are titanium, zirconium, hafnium, vanadium, niobium, tantalum, chromium, molybdenum, tungsten, technetium, tin and antimony. Manganese(II) is very poorly extracted (1–2%). Of iron, cobalt and nickel, only the extraction of iron(III) is of significance (15% from 1 N acid). Those elements which are not extracted are beryllium, magnesium, calcium, strontium, barium, scandium, yttrium, lanthanum, rare earths, thorium, uranium, neptunium, plutonium, ruthenium, zinc, cadmium, mercury, boron, aluminium, gallium, indium, thallium, lead and bismuth.

From 1 N hydrochloric acid, 70% of titanium is extracted and 100% of zirconium and hafnium. At increased acidities, titanium extraction increases to 100% at 8 N, zirconium and hafnium give minima with about 93% and 89% extraction, respectively, at 5 N, but with a further increase in acid concentration extraction efficiency increases to about 100% above 8 N for zirconium and over 10 N for hafnium.

Vanadium(IV) is not extracted at any concentration of acid, whereas for vanadium(V) the per cent extraction increases from

92 from 1 N acid to 97 from 5–6 N acid and then it decreases to 88 from 8 N acid and increases again to 100% from 11 M.

For niobium, the extraction is 94% from 1 N acid increasing to 100% as the acid concentration increases to 10–11 N. For tantalum, the extraction is 80% from 1 N acid. It decreases to 62% from 3 N and again increases to 95% from 6 N acid. From 7 N acid it gives another minimum with 86% extraction. It is completely extracted from 10–11 N acid.

Chromium(III) is not extracted and the extraction efficiency of chromium(VI) decreases from 86% from 1 N acid to 2% from 11 N acid. For molybdenum(VI), the extraction is > 99% from 1–8 N acid and for tungsten, the per cent extraction is 89 from 1 N acid, 95 from 5 N and with a further increase in acidity (> 6 N) the extraction efficiency is constant at 92%.

For technetium, the per cent extraction increases from 8 from 1 N to over 70 from 6·5 N acid and then decreases to about 20 from 11 N acid.

About 90 per cent of antimony(III) and 93% of tin(IV) are extracted from 1 N acid; as the acidity increases their extraction efficiency decreases.

Procedure

Dissolve each element (0·5 or 1·0 mg as the sulphate) in hydrochloric acid (25 ml) of desired normality between 1 and 11 N. Transfer to a separating funnel and add 0·5% BPHA soln. in chloroform (25 ml). After a contact period of 15 min, allow to stand for 5 min for the phases to separate, drain off the phases separately and wherever possible directly determine their metal content by atomic absorption, colorimetric or radiometric methods.

Alternatively, evaporate to dryness the aqueous and organic extracts separately in platinum dishes, digest the residues with nitric and sulphuric acids, fume to dryness, ignite at 500° for 20 min and then analyse by the above methods.

Separation of Niobium and Zirconium from Uranium

In a 200-ml platinum dish dissolve the uranium sample (1–5 g) with 8 N nitric acid. Add to this conc. sulphuric and hydrofluoric acids (10 ml each), fume to dryness and dissolve the residue in 10 N hydrochloric acid (30 ml). Transfer to a separating funnel. Rinse with 10 N hydrochloric acid and add the rinsings to the main soln.

Add a 0·5% chloroformic BPHA soln. (20 ml), shake for 10 min, allow the phases to separate for 5 min and drain off the extract into another separating funnel. Add another 15-ml portion of the BPHA soln., shake for 5 min, combine the extracts and wash the extract twice for 30 sec with 10 N hydrochloric acid (25 ml). Determine niobium and zirconium in the extract by suitable means.

Separation of Cations by Extraction
(Förster and Schwabe[34])

With the use of diethylammonium diethyldithiocarbamate (DADDTC) and BPHA solutions in chloroform as extractants, attempts have been made to separate cations into four main groups. The classical sulphide group is separated by extraction with DADDTC while the ammonia group is separated by extraction with BPHA. The cations of each group are further separated by re-extraction from the chloroform phase with solutions of various reagents. The method of separation is independent of the anions present, and μg and mg quantities of many cations are separated quantitatively.

For extraction into the chloroform phase, having a volume generally twice that of the aqueous phase, a shaking period of 30 sec has been employed. But for re-extraction into the aqueous phase, the period of shaking has been extended to 60 sec. For complete extraction, each process should be repeated 1–2 times with fresh reagents. In all, 50–60 ml of the extractant and 15–25 ml of the re-extractants are used. For larger quantities of the metal ions, the volumes of the extractants should be increased, and the completeness of extraction should be judged by the non-appearance of turbidity on dropping an ethanolic reagent solution into the test solution.

The scheme outlined in Table 6.5 shows the separation of metal ions by extraction into 5 groups with BPHA and DADDTC solutions.

The first group of elements (Table 6.5) are extracted from ~ 2 N mineral acid solution by repeated shaking with an 0·1 M chloroformic BPHA solution. For the extraction of zirconium, hafnium, niobium, tantalum, molybdenum and tungsten, the acid strength of the solution may as well be increased to 10 N (hydrochloric acid) or to 5 N (nitric or sulphuric acids).

TABLE 6.5. EXTRACTIVE SEPARATION OF METAL IONS INTO GROUPS USING BPHA AND DADDTC

Extractant	Acidity of aqueous phase	Metal ions extracted
BPHA (0·1 M)	2 N mineral acid (sulphuric, hydrochloric or nitric)	GROUP I Ti(IV), Zr(IV), Hf(IV), V(V), Nb(V), Ta(V), Mo(VI), W(VI), Sn(II,IV), Sb(III,V), Re(VII)
DADDTC (0·05 M)	2 N mineral acid as above	GROUP II Cu(II), Ag(I), Cd(II), Hg(II), Au(I,III), In(III), Tl(III), Pb(II), As(III), Bi(III), Se(IV), Te(IV), Pd(II), Po
DADDTC-BPHA (0·1 M)	pH 5–8	GROUP III Be(II), Zn(II), Al(III), Ga(III), Sc(III), R.E.(III), Cr(III), Mn(II), Co(II), Ni(II)
BPHA (0·3 M)	1 M ammoniacal	GROUP IV Ca, Sr, Ba
—	—	GROUP V Li, Na, K, Rb and Cs remain in aqueous phase

The extraction of niobium increases with increasing hydrochloric acid concentration above 2 N and is 100% from 10 N acid. The tantalum extraction can be carried out from dilute hydrofluoric acid; > 98% of tantalum is extracted by a single step from a solution 2 N in mineral acid and 0·1 M in hydrofluoric acid.

For tin extraction from hydrochloric and sulphuric acids, the strength of the acids may be increased to 3 N and 5 N, respectively. Rhenium(VII) is well extracted only from hydrochloric acid of higher normality. For instance, rhenium extraction from 10 N hydrochloric acid is 80% and that form 5 N acid is only 1%, whereas from 10 N sulphuric acid it is 22%.

During extraction of the second group of elements from ~ 2 N mineral acid by repeated shaking with a 0·05 M chloroformic DADDTC solution, it is observed that if the acidity of the aqueous solution is more than 2 N, cadmium (from sulphuric acid) and cadmium, lead, indium and thallium (from hydrochloric acid) are only partly extracted. Copper, silver, mercury, arsenic, bismuth, selenium, tellurium and palladium can be extracted completely from 5 N acid. If, however, the acidity of the test solution drops below 1 N, zinc and cobalt have the chance of being extracted along with the second group elements. Therefore, for extracting with DADDTC solution, if its volume is 5 times the volume of the 2 N mineral acid solution, ~ 1 ml of 10 N mineral acid should be added to the test solution for every 100 ml of extractant.

For the details of the extraction of the second main group metals with DADDTC solution only and the subsequent re-extraction for the further separation, the reader may consult the original paper.[34]

For the extraction of the metal ions of the third main group, BPHA and DADDTC (0·5 M in acetone solution) should first be added in excess to the test solution, the pH of which is then adjusted to 5–8 and the precipitate due to chelate formation is afterwards extracted by shaking with pure chloroform. The completeness of separation is determined by the addition of fresh reagents to the aqueous solution. The extraction of chromium(III) is complete only when the precipitate is allowed to stand for a longer period or processed hot. All other cations even in larger amounts do not offer any difficulty.

To extract the fourth main group of metals, the test solution is first made 1 M in carbonate-free ammonia and then repeatedly shaken with a 0·3 M BPHA solution in chloroform. The alkali metals are not extracted and so remain in the aqueous phase.

In the Tables 6.6–6.9 are described the schemes for subsequent back-extraction and separation of metals by different wash solutions.

The chloroform extract of the first group is shaken with 2 N hydrochloric acid, when rhenium is washed to the extent of 98% into the aqueous phase, which may be added to the original solution. Only from concentrated hydrochloric acid is > 99% of rhenium extracted into 0·1 M chloroformic BPHA solution. Three extractions must be made.

By repeated washing of the chloroform extract with 5–10 N sulphuric acid, titanium, zirconium and hafnium are removed as their

TABLE 6.6. BACK-EXTRACTION SEPARATION OF FIRST GROUP OF METAL IONS

Wash solutions	Metal ions re-extracted
2 N HCl	Re
10 N H_2SO_4	Ti, Zr, Hf
0·25 M oxalic acid	Sb, Sn
0·5 M HF + 1 M NaF	Nb
10 N HF	Ta
2 N NaOH	Mo, W

TABLE 6.7. BACK-EXTRACTION SEPARATION OF SECOND GROUP OF METAL IONS

Wash solutions	Metal ions re-extracted
Buffer, pH 9	None
0·2 M KCN	Cu, Ag, Cd, Hg, Se
5 N HCl	Pb, In
10 N HCl	Bi, Tl, Te
0·1 M Na_2S	As
10 N HCl + $KClO_3$	Pd

TABLE 6.8. BACK-EXTRACTION SEPARATION OF THIRD GROUP OF METAL IONS

Wash solutions	Metal ions re-extracted
Buffer, pH 8–9	None
0·2 M KCN	Ni, Zn
0·1 N HCl	Lanthanides, Sc, Mn
10 N HCl	Ga, Mn, Al, Cr
10 N HCl + $KClO_3$	Co

TABLE 6.9. BACK-EXTRACTION SEPARATION OF FOURTH GROUP OF METAL IONS

Wash solutions	Metal ions re-extracted
0·1 N HCl	Ca, Sr, Ba

sulphato complexes; > 90% removal takes place by a single washing with 10 N sulphuric acid. Zirconium and hafnium are extracted together in higher yield. Only 0·5% remain in the aqueous phase and during washing < 2% in the chloroform. About 5% of tin and < 0·5% of antimony accompany zirconium and hafnium in the sulphuric acid washing. By a single shaking with 0·25 M oxalic acid, > 90% of tin and antimony are washed out.

The niobium–BPHA complex is stable towards 5 N sulphuric acid and 0·1 M oxalic acid, whereas the tantalum–BPHA complex is very stable in 25 N sulphuric acid and 0·5 M oxalic acid. About 10% of tantalum accompanies niobium during the washing out of the latter into dilute hydrofluoric acid.

By shaking thrice with 2 N sodium hydroxide solution, molybdenum and tungsten are re-extracted into the aqueous phase. The loss of molybdenum is about 5% during tantalum re-extraction into 10 N hydrofluoric acid, 1·4% in the oxalic acid process and < 1% in other steps.

For the re-extraction of the elements of the third main group, as a first step the excess of DADDTC is washed out by shaking with an ammonium chloride–ammonia buffer of pH 9 and then, by successive washing with 0·2 M potassium cyanide solution and 0·1 N hydrochloric acid, nickel and zinc, and manganese, scandium and the lanthanides, respectively, are recovered. Only after repeated shaking with 10 N hydrochloric acid are chromium, gallium and aluminium washed out. Manganese distributes between the lanthanide and aluminium groups. Re-extraction of cobalt is possible only with 10 N hydrochloric acid containing potassium chlorate.

Calcium, strontium and barium, comprising the fourth main group are washed out by shaking twice with 0·1 N hydrochloric acid.

For the separation of manganese from scandium and lanthanides the hydrochloric acid wash is made 0·01 M in tartaric acid, neutralized with ammonia and shaken with 0·05 M DADDTC solution.

For the separation of gallium, manganese, aluminium and chromium, the 10 N hydrochloric acid solution containing these elements is diluted to reduce the acidity to 5–6 N and then the gallium is extracted by repeated shaking with ether. The aqueous phase is made 0·01 M in tartaric acid and after adjusting its pH to

7–9 is shaken with 0·05 M chloroformic DADDTC solution to extract manganese. In the same way and at the same pH, aluminium and chromium are extracted sequentially with an 0·1 M chloroformic BPHA solution. Chromium is extracted as its BPHA complex after the addition of fresh reagent and boiling for 10 minutes (Table 6.10).

TABLE 6.10. BACK-EXTRACTIVE SEPARATION OF GALLIUM, MANGANESE, ALUMINIUM AND CHROMIUM

Extractant	Acidity of the aqueous phase	Metal ions extracted
Ether	6 N hydrochloric acid	Ga
DADDTC (0·05 M)	pH 7–9	Mn
BPHA (0·1 M)	pH 7–9	Al
BPHA (0·1 M)	pH 7–9 (boil)	Cr

Extraction of Germanium and Separation from Gallium
(Alimarin, Sokolova and Smolina[35])

The extraction of the germanium–BPHA complex into chloroform is complete when the acidity of the aqueous phase is $\geqslant$ 2 N in perchloric acid, $\geqslant$ 3 N in hydrochloric acid or $\geqslant$ 12 N in sulphuric acid and the aqueous to organic phase volume ratio is 1:1. As the gallium–BPHA complex is extracted into chloroform from an aqueous phase at pH 2–9, germanium can be separated from 10,000-fold excesses of gallium either by extraction of germanium from 5 M hydrochloric acid solution, leaving gallium in the aqueous phase, or by the extraction of gallium from hydrochloric acid at pH 3–5, leaving germanium in the aqueous phase.

Procedure

For the extraction of germanium (5–20 μg), add enough conc. hydrochloric acid to make the final 10-ml of soln. 5 N in acid, followed by either a 5% ethanolic BPHA soln. (0·2 ml), water to dilute to 10 ml and chloroform (10 ml) or water to dilute to 10 ml and a 0·1% chloroformic BPHA soln. (10 ml). Shake for 3 min to extract. For re-extraction into an aqueous phase, twice shake the organic phase with 10-ml portions of water, each time for 5 min.

References

1. SHENDRIKAR, A. D., *Talanta* **16**, 51 (1969).
2. DYRSSEN, D., *Acta Chem. Scand.* **10**, 353 (1956).
3. ZHAROVSKII, F. G., *Ukrain. Khim. Zhur.* **25**, 245 (1959).
4. ALIMARIN, I. P. and YUN-HSIANG, CHIEH, *Zavod. Lab.* **12**, 1435 (1959); *Talanta* **8**, 317 (1961).
5. VILLARREAL, R., KRSUL, J. R. and BARKER, S. A., *Anal. Chem.* **41**, 1420 (1969).
6. ALIMARIN, I. P. and YUN-HSIANG, CHIEH, *Vestnik Moscow Univ., Khim.* **15**, 53 (1960).
7. CHE-MING, NI, CHUNG-FEN, CHU and SHU-CHUAN, LIANG, *Acta Chim. Sinica* **29**, 249 (1963); *C.A.* **62**, 5872 (1965).
8. CHE-MING, NI, CHUNG-FEN, CHU and SHU-CHUAN, LIANG, *Hua Hsueh Hsueh Pao* **30**, 290 (1964); *C.A.* **61**, 11309 (1964).
9. HALA, J., *J. Inorg. Nucl. Chem.* **29**, 187 (1967).
10. FOUCHE, K. F., *Talanta* **15**, 1295 (1968).
11. FOUCHE, K. F., *J. Inorg. Nucl. Chem.* **30**, 3057 (1968).
12. ALIMARIN, I. P., PETRUKHIN, O. M. and YUN-HSIANG, CHIEH, *Dokl. Acad. Nauk SSSR* **136**, 1073 (1961).
13. MAJUMDAR, A. K. and MUKHERJEE, A. K., *Anal. Chim. Acta* **21**, 245 (1959).
14. ALIMARIN, I. P. and PETRUKHIN, O. M., *Russian J. Inorg. Chem.* **7**, 612 (1962).
15. ALIMARIN, I. P. and PETRUKHIN, O. M., *Analytical Chemistry (Proc. Intern. Symp. Birmingham Univ., England)* **1962**, p. 152 (pub. 1963).
16. PALSHIN, E. S., MYASOEDOV, B. F. and NOVIKOV, YU. P., *Zhur. Anal. Khim.* **18**, 657 (1963).
17. MYASOEDOV, B. F., PALSHIN, E. S. and PALEI, P. N., *Zhur. Anal. Khim.* **19**(1), 105 (1964).
18. LYLE, S. J. and SHENDRIKAR, A. D., *Radiochim. Acta* **3**, 90 (1964).
19. RAKOVSKII, E. E. and PETRUKHIN, O. M., *Zhur. Anal. Khim.* **18**, 539 (1963).
20. LYLE, S. J. and SHENDRIKAR, A. D., *Anal. Chim. Acta* **32**, 575 (1965).
21. LYLE, S. J. and SHENDRIKAR, A. D., *Talanta* **12**, 573 (1965).
22. ERSKINE, J. S., SINK, M. L. and VARGA, L. P., *Anal. Chem.* **41**, 70 (1969).
23. CHMUTOVA, M. K., PETRUKHIN, O. M. and ZOLOTOV, Yu. A., *Zhur. Anal. Khim.* **18**, 588 (1963).
24. LYLE, S. J. and SHENDRIKAR, A. D., *Talanta* **13**, 140 (1966).
25. LYLE, S. J. and SHENDRIKAR, A. D., *Anal. Chim. Acta* **36**, 286 (1966).
26. SCHWEITZER, G. K. and NORTON, A. D., *Anal. Chim. Acta* **30**, 119 (1964).
27. CHWASTOWSKA, J. and BRAGINSKA, J., *Chem. Anal. (Warsaw)* **11**, 169 (1966).
28. CHWASTOWSKA, J. and MINCZEWSKI, J., *Chem. Anal. (Warsaw)* **8**, 157 (1963).
29. CHWASTOWSKA, J. and MINCZEWSKI, J., *Chem. Anal. (Warsaw)* **9**, 791 (1964).
30. CHWASTOWSKA, J., *Chem. Anal. (Warsaw)* **12**, 469 (1967).
31. CHWASTOWSKA, J., *Proc. Conf. Appl. Phys. Chem. Methods Chem. Anal.*, Budapest, **1**, 45 (1966); *C.A.* **68**, 56251 (1968).
32. MAREVA, S., PUZDRENKOVA, I. V. and STOYANOVA, T., *Bulg. Akad. Nauk* **5**, 5 (1967); *C.A.* **69**, 15713 (1968).
33. VITA, O. A., LEVIER, W. A. and LITTERAL, E., *Anal. Chim. Acta* **42**, 87 (1968).
34. FÖRSTER, H. and SCHWABE, K., *Anal. Chim. Acta* **45**, 511 (1969).
35. ALIMARIN, I. P., SOKOLOVA, I. V. and SMOLINA, E. V., *Vestnik Moscow Univ., Khim.* **23**, 67 (1968).

CHAPTER 7

TITRIMETRIC AND PAPER CHROMATOGRAPHIC APPLICATIONS OF N-BENZOYLPHENYLHYDROXYLAMINE

The colour-producing character of N-benzoylphenylhydroxylamine (BPHA) as a chelating agent finds application as an indicator in acetone or ethanol solution or in solution in chloroform for more precise extractive end-point determinations in complexometric titrations. The iron(III)–BPHA complex is also used as a metallochromic indicator for the EDTA titration of metal ions.

For the direct titration with an ethanolic BPHA solution, the end-point is detected by iron(III) or amperometrically either on the basis of the reduction current at a dropping mercury electrode or on the basis of the oxidation current at a rotating platinum or a graphite micro-electrode. In cases where the oxidation or reduction current is very small, the excess BPHA is titrated with an oxidant or, at a dropping mercury electrode, with a metal ion solution that precipitates the excess BPHA.

Studies of the chromatographic behaviour of a large number of metal ions on papers impregnated with BPHA has enabled several mixtures of metal ions to be separated, depending upon the stability and nature of their respective BPHA complexes. These final examples depict yet another facet of the application of BPHA as an analytical reagent.

Determination of Iron
(Alimarin and Yun-hsiang[1])

Iron(III) reacts stoichiometrically with EDTA at pH 1·0–1·5 when the temperature of the solution is kept at 50–60° and therefore can be titrated compleximetrically under these conditions in the presence of 1 ml of a 0·5% acetone solution of BPHA as indicator. The colour change at the end-point is from red-violet to lemon yellow or colourless. The end-point may be detected visually, or spectrophotometrically at 500 nm. Small amounts of cobalt, nickel, copper, aluminium, uranium and the lanthanides do not interfere. The method gives good results for the determination of iron in iron ores and in magnesite bricks.

Determination of Scandium and Zirconium
(Yun-hsiang[2])

For the determination of scandium and zirconium, an excess of EDTA is added to complex the elements and the excess is back titrated with an iron(III) solution under the conditions,[1] described immediately above, using BPHA as indicator. Aluminium, uranium, the lanthanides and beryllium do not interfere.

Determination of Vanadium(V) in the Presence of Fluoride, Phosphate, Titanium, Molybdenum and Manganese
(Kaimal and Shome[3])

Vanadium(IV) forms a blue anionic complex with EDTA in weakly acidic media; its stability constant (log K) is 18·77. With BPHA as indicator, vanadium(IV) can be titrated directly with EDTA. The flesh-coloured complex formed by vanadium(IV) with BPHA at pH 3 becomes red in 50% ethanol; on titration with EDTA the colour change at the end-point is distinct, from red to blue. The titration must be done at room temperature, as at higher temperatures the vanadium(IV)–BPHA complex is unstable. The optimal pH is 2·5–4·5.

Procedure

To a vanadium(V) soln. (containing 3–38 mg of the metal) add dilute sulphuric acid and sodium sulphite as required. Heat to reduce the vanadium to vanadium(IV). Boil to remove the excess of sulphur dioxide. Add a sodium acetate–acetic acid buffer soln. to adjust the pH to 3. Dilute to 100 ml with 50% ethanol and add with stirring a 2% soln. of BPHA in 95% ethanol (5 ml); the soln. becomes red. Titrate with a standard 0·01 M EDTA soln. till a permanent blue colour appears at the end-point.

Effect of Other Ions

In the titration of 9·6 mg of vanadium at pH 4·5, titanium(IV) ($\leqslant$ 23·5 mg) and molybdenum(VI) ($\leqslant$ 10 mg) as ammonium molybdate can be present. Vanadium (7·5–15·0 mg) can be determined in the presence of 2–10 mg of manganese(II) and up to 150 and 200 mg of potassium fluoride and diammonium hydrogen phosphate, respectively. Iron(III), if present, must be removed before the determination of vanadium. To do this, add to the soln. containing iron (2·8–14·0 mg), present as its sulphate, a few drops of dilute sulphuric acid and a 20-volume hydrogen peroxide soln. (2–3 ml). Dilute, and precipitate iron as its hydroxide by adding dilute ammonia soln. Filter, wash, evaporate the soln. to a very small volume, acidify with dilute sulphuric acid, boil with sodium sulphite and titrate vanadium(IV) as described above.

Determination of Iron and Copper in the Presence of Many Other Ions by Using an Extractive End-point Procedure

(Das and Shome[4])

Iron(III) and copper form reddish-violet and greenish-yellow complexes, respectively, with BPHA which are extractable into chloroform. During the titration of iron(III) and copper with EDTA using BPHA as the indicator, the end-points are judged from the colour of the chloroform extracts which change from red for iron and greenish-yellow for copper to colourless. Iron(III) can also be titrated with the use of an alcoholic BPHA solution as indicator. The pH of the solutions adjusted by dilute hydrochloric acid and/or sodium acetate must be between 0·8–1·4 for iron and 3·9–5·4 for copper. The pH of the iron(III) solution can be maintained at 1·5–2·0 when the indicator is alcoholic BPHA soln. For the titration of iron, although the best results are obtained with the use of 6–8 ml of 0·0008 M to 0·002 M BPHA in chloroform, BPHA just sufficient to colour the chloroform layer can also be used. A more than 0·003 M BPHA soln. tends to give erratic results. When, on the other hand, an alcoholic solution of BPHA is used as the indicator, variations in the concentration of BPHA do not have much effect. The extraction procedure gives a sharper end-point. For the titration of copper, the use of 10–15 ml of an 0·005 M BPHA solution in chloroform is best. Slight variations in the concentration of BPHA does not affect the results, but a large excess of BPHA must be avoided.

Determination of Iron(III)

Procedure (a)

Acidify the iron(III) soln. (containing 3–14 mg of iron) diluted to 50 ml (10 ml for 0·340–0·085 mg Fe) with dilute hydrochloric acid to give a pH of 1·0–1·3. Add an 0·001 M BPHA soln. in chloroform (8–10 ml) and titrate with a standard EDTA soln. Use 0·001 M and 0·0001 M EDTA solns. for the mg and sub-mg amounts of iron, respectively. Towards the end-point, shake vigorously after each addition of EDTA. At the end-point, the chloroform layer becomes colourless.

Procedure (b)

Dilute the iron(III) soln. to 100 ml (10 ml for 0·170–0·085 mg Fe) and acidify with dilute hydrochloric acid to adjust the pH to 1·5–2·0. Add an alcoholic 1% BPHA soln. (10–12 drops) and titrate with standard EDTA soln. till at the end-point the soln. changes to colourless or pale yellow from red-violet.

Determination of Copper

Adjust the pH of the copper sulphate soln. (containing 4–56 mg of copper) diluted to 50 ml to 4·2–5·4 by the addition of a 10% sodium acetate soln. and dilute hydrochloric acid. Add a 0·005 M BPHA soln. in chloroform (10–15 ml) and titrate with standard 0·01 M EDTA soln. till at the end-point the chloroform layer becomes colourless.

Effect of Other Ions

Iron (0·341 mg) can be determined by procedure (a) at pH 1·2 with the use of 10–15 ml of 0·001 M BPHA in chloroform, as indicator, in the presence of cadmium, mercury(II), uranium(VI), cerium(III), manganese(II), aluminium, calcium and magnesium and by procedure (b) at pH 1·5 in the presence of zinc, chromium(III), nickel and all the ions mentioned above. Molybdenum(VI), thorium and vanadium(V) interfere in both methods. The interfering effect of copper and nickel in procedure (a) is, however, masked by tartrate. Molybdenum(VI) (≤115 mg) and titanium(IV) (6–7 mg) have no effect on the titration of iron (0·341 mg) in the 50% ethanolic soln. using 1% BPHA in ethanol as indicator.

By the extraction procedure, copper (4·65 mg) can be titrated at pH 4·4–4·6 in the presence of zinc, manganese(II), chromium(III), molybdebum(VI), mercury (II), uranium(VI) and cerium(III). The chloroform layer must be made distinctly greenish yellow by the addition of more indicator for the titration of copper in the presence of larger amounts (50 mg) of molybdenum(VI). The interfering effects of iron(III), titanium(IV), thorium and aluminium are masked by the addition of fluoride before the titration of the copper.

Determination of Indium, Thallium and Thorium with Iron–BPHA as a Metallochromic Indicator

(Das and Shome[5])

The strongly coloured iron(III)–BPHA complex has been used as a metallochromic indicator in the EDTA titration of small amounts of indium, thallium(III) and thorium at pH 1·5–6·0. The optimum pH range, however, is 2–4. The colour change at the end-point, reddish-violet to colourless, is very sharp. The upper limits for the titrimetric determination of indium and thallium are 2·5 and 9·2 mg, respectively. Thorium, however, can be determined even when more than 35 mg is present.

Indicator: 1% ethanolic BPHA soln. (5 ml) and a 0·001 M iron(III) soln. (2·5 ml) are used for each titration.

Procedure

Dilute the soln. containing indium, thallium or thorium to 50 ml in a 200-ml conical flask. Adjust the pH to 2–3 by adding dilute hydrochloric acid or a suitable quantity of 10% sodium acetate soln. Add the indicator. If precipitate appears, add ethanol (10–15 ml) to dissolve it. Titrate with a standard 0·01 M EDTA soln. to a disappearance of the colour at the end-point. Deduct 0·2 ml from the burette reading as the indicator correction.

Effect of Other Ions

By the above procedure, thorium(IV) (5·30 mg), thallium(III) (2·33 mg) and indium(III) (2·16 mg) can be titrated with EDTA at pH 2 in the presence of aluminium (50 mg), chromium(III) (8 mg), manganese(II) (25 mg), zinc (100 mg), uranium(VI) (12 mg), nickel (7 mg), mercury(II) (150 mg), cerium(III) (45 mg), calcium (100 mg), or magnesium (120 mg). Titanium(IV), vanadium(V) and molybdenum(VI) interfere when present even in traces and so must be absent.

Notes

1. For a titration in the presence of nickel, the pH must be adjusted to 1·5–1·8, because the nickel–EDTA complex is stable at a higher pH.

2. For a titration in the presence of chromium(III), the colour change at the end-point is from violet to light green.

N-Benzoylphenylhydroxylamine as a Titrant for the Determination of Zirconium

(Zharovskii, Shpak and Piskunova[6])

Zirconium forms a 1:4 complex with BPHA, which in solution in chloroform is colourless. The BPHA–iron(III) complex, however, is red in a chloroform solution. As the zirconium complex is stronger and can be extracted into chloroform from 2 N hydrochloric or sulphuric acid, an extractive titrimetric method has been developed for the determination of zirconium in 2 N acid with iron(III) as the indicator. At the end-point, when all the zirconium has been extracted, the iron(III)–BPHA complex forms, giving a pink colour to the chloroform extract. The error in the determination of 1·9 mg of zirconium is within $\pm 1\%$. The method is satisfactory for the rapid determination of zirconium in magnesium alloys and in zirconium oxide. Iron, aluminium, zinc, manganese, lead, copper, bismuth and the alkali and alkaline earth metals do not interfere. Titanium, niobium, tantalum, molybdenum, tungsten and vanadium must be absent.

Determination of Zirconium

Adjust the acidity of the zirconium soln. to 2 N by adding dilute hydrochloric or sulphuric acid. Add chloroform (10 ml) and a 0·1 M iron(III) soln. (2–3 drops). Titrate with a 50% aqueous ethanolic soln. 0·01 M in BPHA, shaking vigorously towards the end-point, until at the end-point the chloroform layer becomes pink.

Amperometric Determination of Titanium, Zirconium, Gallium, Scandium and Other Elements

(Gallai, Alimarin and Sheina[7])

Current–voltage measurements made with a dropping mercury electrode show that though the reduction wave of BPHA fuses with that of hydrogen ions at −1·0 V in 0·1–3·0 N sulphuric acid solution, there is a proportionality between the concentration and the current. This suggests that an amperometric titration can be undertaken on the basis of the reduction current of the reagent. Moreover, BPHA has the added advantage that the end-point can also be detected on the basis of its oxidation current at a rotating platinum electrode.

Determination of Titanium

With a 0·1 M ethanolic BPHA soln., titanium(IV) (0·7–4·0 mg per 10 ml of solution) is determined by titration using a rotating platinum electrode at 1 V. Results are most reproducible in a supporting electrolyte of 1 N sulphuric acid. Iron(III) ($\not>$ 30 times the amount of titanium) must be masked with EDTA. Up to a 100-fold excess of nickel, manganese, cobalt, aluminium or chromium does not interfere. The method has been used for the determination of titanium in alloys containing 2·4% of titanium.

Procedure

Dissolve the alloy (1 g) in (1 + 1) hydrochloric acid, transfer the soln. to a 100-ml graduated flask and make up to volume. Neutralize an aliquot (5 ml) of the soln. with potassium hydroxide soln. till the soln. is turbid. Add 0·5 N sulphuric acid (10 ml), a 0·15 M disodium EDTA soln. (1 ml) and titrate with the BPHA soln. using a rotating platinum electrode as described above.

Determination of Zirconium

Zirconium (0·5–5·0 mg in 10 ml of solution) in a supporting electrolyte of 1 N sulphuric acid can be determined under the conditions described for titanium using a rotating platinum electrode. Nickel, chromium, manganese, cobalt and aluminium do not interfere even when present to the extent of about a 100-fold excess. A dropping mercury electrode can also be used. Iron(III), if reduced by ascorbic acid, can be tolerated to the extent of 100 times the amount of zirconium.

Determination of Gallium

Gallium (0·4–6·0 mg per 10 ml) in a supporting electrolyte of hydrochloric acid–potassium hydrogen phthalate buffer can be determined using a rotating platinum electrode in the presence of aluminium (65-fold excess) at pH 2·4–3·0, zinc or manganese ($\not>$ 500-fold excess) at pH 2·4–4·0, and indium (100-fold excess) at pH 2·4. In the presence of lead (500-fold excess) gallium can be determined at pH 3–4 in an acetate–ammonia supporting electrolyte.

Iron(III) interferes unless reduced by ascorbic acid, but as iron(II) gives an oxidation current at the titration potential at the rotating platinum electrode, the determination of gallium, in its presence, is performed with a dropping mercury electrode at $-1\cdot2$ V in a supporting electrolyte of 0·01 N hydrochloric acid adjusted to a pH $\geqslant 2\cdot4$ by dilute alkali. Satisfactory results are then obtained in the presence of a 500-fold excess of iron.

Determination of Scandium

Scandium is completely precipitated by BPHA at pH 5·4, but because under this condition the oxidation or reduction current of BPHA is very small, a 0·1 N ethanolic BPHA solution is added to the test solution in excess and the excess is oxidized by back titration with a 0·2 N potassium permanganate solution. This procedure gives satisfactory results for 0·2–1·5 mg of scandium in 10 ml of solution.

Using a dropping mercury electrode, back titration of the excess BPHA can also be carried out with a zinc salt solution of known concentration at a potential of $-1\cdot4$ V, maintaining the pH for zinc precipitation at 6. This allows the determination of 0·4–4·0 mg of scandium in 10 ml of solution.

Copper sulphate solution containing 2 mg of copper per ml has also been used for the back titration of the excess BPHA, at a potential of $-0\cdot6$ V. The pH is kept at 5·5 with a buffered supporting electrolyte. Because the lanthanides are not precipitated by BPHA at this pH, scandium is determined in presence of 100 times its weight of neodymium, lanthanum and yttrium oxides.

Determination of Other Elements

Amperometric titrations with a BPHA solution can also be

undertaken to determine molybdenum(VI) and copper at pH 4 using a platinum electrode and zinc, cadmium, nickel and lead at pH 6 using a dropping mercury electrode.

Amperometric Determination of Gallium in Arsenides and Phosphides

(Gallai, Sheina and Alimarin[8])

BPHA is oxidized in solutions 1–18 N in sulphuric acid or at pH 3·5–8·5 at a graphite microelectrode. In 1 N sulphuric acid, the half-wave potential is 0·9 V (vs. S.C.E.). This potential decreases with increasing pH. In acidic and basic electrolytes, the diffusion current is proportional to the concentration of BPHA. On the basis of the oxidation current of BPHA at 1·1 V, gallium can be determined by titration at pH 3. Arsenic and phosphorus, present in amounts 1·5 times that of gallium, do not interfere.

Procedure

Weigh out accurately the powdered test material (5 mg) into a micro-beaker and add conc. sulphuric acid (1 ml) and powdered ammonium sulphate (0·15 g). Heat to decompose. After cooling, add water (2 ml) and heat again to dissolve the gallium sulphate. Transfer the soln. to a 25-ml graduated flask, neutralize with sodium hydroxide soln. to the yellow colour of methyl orange and add sulphuric acid dropwise until the colour becomes rose. Dilute to the mark with water. Place an aliquot (2 ml) of the soln. in a micro-cell and titrate the gallium with an 0·1 M ethanolic BPHA soln. from a 0·2-ml micro-burette, using a graphite electrode at 1·1 V (vs. S.C.E.). Standardize the BPHA soln. amperometrically with a standard gallium soln.

Amperometric Determination of Niobium

(Gallai, Sheina and Nifontova[9])

With the use of a graphite indicator electrode at 0·95 V (vs. S.C.E.), an attempt to amperometrically determine niobium in a strongly acidic medium has been made with BPHA as the titrant. The latter is oxidized at the solid micro-electrode at a wide range of acidities and the anodic current can be utilized to determine the end-point. As supporting electrolytes, 1–5 N sulphuric acid or 1–9 N hydrochloric acid do not give good results, perhaps because of the tendency of niobium to form polymeric compounds. However, on titration in 10–12 N hydrochloric acid as the supporting electrolyte, niobium and BPHA appear to bear a constant ratio of 1:4 at the end-point.

On the addition of first few ml of the reagent solution, the current increases steeply and the solution turns yellow. This is accompanied by the formation of a precipitate of the electroactive niobium–BPHA complex. The second portion of the titration curve with a small slope corresponds to the oxidation current of the excess reagent.

Satisfactory results are obtained over a wide range of niobium concentration (0·3–4·0 mg) on titrating in 10·2 N hydrochloric acid as supporting electrolyte. At such a high acidity niobium (1 mg) can be determined in presence of 500 times its weight of nickel, cobalt, zinc, aluminium, manganese(II), chromium(III) and iron(III), 100 times its weight of vanadium(IV) and 9 times its weight of rhenium(VII). Zirconium a 10-fold excess in 10 ml of solution does not interfere in the determination of niobium (0·5 mg). Tantalum, titanium, molybdenum and tungsten interfere.

The method has been used for the determination of niobium in niobium–rhenium and niobium–zirconium–rhenium alloys, containing about 80% and 40% of niobium, respectively.

Procedure

Dissolve the alloy (0·2 g) in a mixture of conc. hydrofluoric acid (20 ml) and conc. sulphuric acid (1 ml) in a platinum dish, also, adding nitric acid (1–2 drops). Evaporate to 4 ml and transfer to a 100-ml graduated flask using a 5% tartaric acid soln. as a wash liquid and to make up to volume. Take an aliquot (1 ml) and titrate with an 0·1 M ethanolic BPHA soln. which has been standardized against a standard niobium soln., prepared from spectrographically pure niobium pentoxide in the same way as above, but without the addition of the nitric acid and containing 1 mg of niobium per ml.

Paper Chromatographic Separation of Metal Ions
(Fritz and Sherma[10])

Papers impregnated with BPHA have been used for the study of the chromatographic behaviour of thirty-five different metal ions using

TABLE 7.1. SEPARATION ATTEMPTS BY CHROMATOGRAPHY ON BPHA-IMPREGNATED PAPER

Developer	Pairs of metals (with R_f values below) for trailing and leading edges				
1 M HClO$_4$	Titanium 0·0, 0·0	and	Uranium 0·70, 1·0		
	Zirconium 0·0, 0·0	and	Thorium 0·90, 1·0		
	Vanadium 0·0, 0·0	and	Bismuth 0·58, 0·70		
	Zirconium 0·0, 0·0	and	Aluminium 0·85, 1·0		
	Iron 0·28, 0·68	and	Mercury 0·71, 0·85		
3 M HClO$_4$	Lead 0·60, 0·74	and	Copper 0·89, 1·0		
	Lead 0·60, 0·74	and	Uranium 0·85, 1·0		
5 M HClO$_4$	Bismuth 0·50, 0·72	and	Iron 0·85, 1·0		
HClO$_4$ (pH 2)	Zirconium 0·0, 0·0	and	Aluminium 0·74, 0·92		
	Mercury 0·0, 0·0	and	Cobalt 0·84, 0·92		
HClO$_4$ (pH 3)	Gold 0·0, 0·50	and	Platinum 0·88, 0·97		
	Silver 0·0, 0·75	and	Platinum 0·85, 0·99		
	Tin 0·0, 0·21	and	Arsenic 0·76, 0·87		
	Lead 0·22, 0·55	and	Platinum 0·90, 1·0		
1 M HCl	Silver 0·0, 0·0	and	Gold 0·43, 0·54	and	Copper 0·87, 1·0
	Tin 0·0, 0·0	and	Iron 0·85, 0·98		
	Tin 0·0, 0·0	and	Lead 0·61, 0·81		
	Silver 0·0, 0·0	and	Mercury 0·86, 0·98		
	Gold 0·42, 0·57	and	Mercury 0·86, 0·98		
	Tin 0·0, 0·0	and	Mercury 0·84, 0·99		
	Silver 0·0, 0·0	and	Lead 0·60, 0·80		
6 M HCl	Gold 0·31, 0·51	and	Lead 0·85, 0·98		

developing liquids with acidities ranging from 6 M hydrochloric acid or 5 M perchloric acid to pH 3. Knowing the R_f values of the leading and trailing edges of the spots, a few separations have been achieved. Table 7.1 gives the separations attempted.

Procedure

Impregnate sheets (28 × 40 cm) of Whatman No. 1 chromatography paper or Whatman No. 3 MM paper, by dipping them in a 2% soln. of BPHA in 40% 2-octanone and 60% absolute ethanol and allow the excess soln. to drip away. Dry the sheets for 1 hr.

Prepare developing liquids of proper acidity or pH by adding perchloric acid or hydrochloric acid and sodium hydroxide solns. to distilled water. Pour 2 litres of the developer into a large round glass jar lined with Whatman No. 3 MM paper and covered with a glass cover. Take out the paper after at least 1 hr of saturation.

Spot about 5 µl of each test soln. (15 µl for cadmium, arsenic and aluminium) at a premarked position on the impregnated paper and allow the zones to dry. Stand the paper in the form of a cylinder in the jar with the zones at the bottom. After 2 hr, when the wash liquid has risen 20 cm above the origin, take out the paper, dry it overnight and spray with the appropriate reagent as given below.

For silver, lead, nickel, cadmium, iron, copper, tin, cobalt, bismuth, mercury, vanadium, gold, platinum, arsenic and antimony, spray with yellow ammonium sulphide soln. followed by 6 N hydrochloric acid. For uranium, spray with a 1% potassium hexacyanoferrate(II) soln. followed by hydrochloric acid. For scandium, yttrium, lanthanum, thorium, zirconium, aluminium and the lanthanides, put in an atmosphere of ammonia for 10 min, spray with an alcoholic alizarin soln. followed by 1 N acetic acid and for aluminium, spray with an aluminon soln. followed by exposure to ammonia.

References

1. ALIMARIN, I. P. and YUN-HSIANG, CHIEH, *Vestnik Moscow Univ., Khim.* **16,** 59 (1961); *C.A.* **56,** 10901 (1962).
2. YUN-HSIANG, CHIEH, Candidate's Thesis, Moscow State University, 1960; *Uspekhi Khim.* **31,** 989 (1962).
3. KAIMAL, V. R. M. and SHOME, S. C., *Anal. Chim. Acta* **27,** 594 (1962).
4. DAS, H. R. and SHOME, S. C., *Anal. Chim. Acta* **35,** 256 (1966).
5. DAS, H. R. and SHOME, S. C., *Anal. Chim. Acta* **43,** 140 (1968).
6. ZHAROVSKII, F. G., SHPAK, E. A. and PISKUNOVA, E. V., *Proc. Kiev. Univ.* No. 1, 118 (1962).
7. GALLAI, Z. A., ALIMARIN, I. P. and SHEINA, N. M., *Zhur. Anal. Khim.* **18,** 1442 (1963).

8. GALLAI, Z. A., SHEINA, N. M. and ALIMARIN, I. P., *Zhur. Anal. Khim.* **20**, 1093 (1965).
9. GALLAI, Z. A., SHEINA, N. M. and NIFONTOVA, N. V., *Zhur. Anal. Khim.* **23**, 942 (1968).
10. FRITZ, J. S. and SHERMA, J., *J. Chromatog* **25**, 153 (1966).

INDEX

N-Acetylphenylhydroxylamine
  preparation 39
  spectrophotometric reagent 134
N-Acetylsalicylphenylhydroxylamine
  preparation 46
  spectrophotometric reagent 153
Acid dissociation constants of hydroxylamines 44
Aluminium extraction from complex mixtures 163
Anthranilohydroxamic acid 21
Arsenic, antimony, bismuth and tin separation 179
N-Arylhydroxylamines as spectrophotometric reagents 154
Atomic groupings and substituents, effect of, on reagents 7–29

Benzohydroxamic acid 16
N-Benzoyl-p-chlorophenylhydroxylamine, preparation 43
  spectrophotometric reagent 143
N-Benzoylmethylhydroxylamine
  preparation 46
  spectrophotometric reagent 135
N-Benzoylnaphthylhydroxylamine 36
N-Benzoylphenylhydroxylamine see BPHA
N-Benzoyl-m-tolylhydroxylamine 42
N-Benzoyl-o-tolylhydroxylamine
  applications 100
  preparation 42
  spectrophotometric reagent 142, 144
N-Benzoyl-p-tolylhydroxylamine
  preparation 42
  spectrophotometric reagent 141
Beryllium
  extraction and determination 183
  gravimetric determination 70
Bismuth
  gravimetric determination 83
  separation from lead by BPHA extraction 180
BPHA
  amperometric determinations
    gallium 198, 201
    niobium 201
    scandium 198
    titanium 198
    zirconium 198
  gravimetric determinations
    beryllium 70
    bismuth 83
    cobalt and nickel 61
    copper, iron and aluminium 55, 58
    gallium 77
    germanium 97
    indium 74
    lanthanum 80
    magnesium 97
    mercury 82
    molybdenum 69
    niobium 90
    scandium 60
    tantalum 88
    tin 59
    titanium 85
    thorium 62
    tungsten 78
    uranium 72
    zirconium 67

BPHA *(cont.)*
  extractive determinations and separations
    aluminium 163
    beryllium 183
    bismuth 180, 182
    cations 186
    copper 181
    gallium 172
    germanium 191
    iron 161
    manganese 182
    niobium 166, 167, 170, 176, 184
    plutonium 176
    protactinium 169, 173, 178
    scandium 162
    thallium 180
    thorium 159, 164
    tin 171
    titanium 122, 128
    tungsten 164
    zinc 182
    zirconium 165
  paper chromatographic separation
    metal ions 202
  pH of interaction and extractability 35
  preparation 30
  properties 31
  reactions with metal ions 34, 50
  spectrophotometric determinations
    cerium 133
    iron 121
    mercury 132
    niobium 129–132
    titanium 122–129
    vanadium 109–121
  solubility
    in ethanol-water 31
    in organic solvents 32
  stability constants of chelates 49
  titrimetric determinations
    copper 195
    indium 197
    iron 194, 195
    scandium 194
    thallium 197
    vanadium 194
    zirconium 194, 198

Cations, separation by extraction with BPHA 186
Cerium, spectrophotometric determination 133
N-Cinnamoylphenylhydroxylamine
  applications 99
  preparation 42
  spectrophotometric reagent 138
Cinnamylhydroxamic acid 20
Copper
  gravimetric determination 103
  iron and aluminium, gravimetric determination 55
  iron, lead, nickel, cobalt and cadmium extraction 181
Cupferrons
  effect of atomic groupings and substituents 7–9
  2-fluorenyl 9
  p-phenyl 9

3,5-Dinitrobenzoylphenylhydroxylamine 23
Dissociation constants of hydroxylamines 44
N-Disubstituted hydroxylamines 35

o-Ethoxybenzoylphenylhydroxylamine 23

N-Furoylphenylhydroxylamine 23
  spectrophotometric reagent 150

Gallium
  amperometric determination 200, 201
  extractive determination 172
  gravimetric determination 77
Germanium
  extractive determination 191
  gravimetric determination 97

Hydroxamic acids, effect of atomic groupings and substituents 13–22
Hydroxylamines
  acid dissociation constants 44
  effect of atomic groupings and substituents 22–27

INDEX

Hydroxytriazines, effect of atomic groupings and substituents 10

Indium and gallium, gravimetric determination 74
Indium, thallium and thorium, titrimetric determination 197
o-Iodobenzoylphenylhydroxylamine 23
Iron
  in silicate materials, spectrophotometric determination 121
  in steel and iron ore, gravimetric determination 104
  spectrophotometric determination 121, 134
  titrimetric determination 194
  vanadium and uranium, successive extraction and spectrophotometric determination 139
1:3 Iron(III)–N-acylphenylhydroxylamine complexes
  molar extinction coefficients 39
  stability constants 39

Lanthanum, gravimetric determination 80

Magnesium, gravimetric determination 97
Mercury
  gravimetric determination 82
  spectrophotometric determination 132
Metal ions
  extraction 183
  paper chromatographic separation 202
Metal-N-salicylphenylhydroxylamine chelates 46
Molybdenum, gravimetric determination 69

2-Naphthohydroxamic acid 22
Naphthylhydroxylamine, N-benzoyl derivative 22
Neocupferron 9
N-Nicotinylphenylhydroxylamine preparation 35
Niobium determination
  amperometric 201
  as niobium-BPHA-thiocyanate complex 129
  gravimetric 92
  spectrophotometric 130
Niobium and tantalum
  gravimetric determination 90, 93, 96, 99, 100
  separation 166
  of fluorcomplexes 176
Niobium and zirconium
  extraction 184
  separation 170
Niobium, tantalum, and titanium, gravimetric determination 94
Niobium, tantalum, titanium, zirconium and vanadium, extraction from sulphuric acid 167

Organic reagents in inorganic analysis 1–6
Oxalohydroxamic acid 20

pH range for precipitation of some chelates 46
Phenylacetylhydroxamic acid 20
N-Phenylacetylphenylhydroxylamine, preparation 43
Plutonium, separation by solvent extraction 176
Protactinium separation from
  neutron-irradiated thorium 178
  niobium, tantalum and hafnium 169, 173

Quinaldinohydroxamic acid 21

Salicylhydroxamic acid 18
N-Salicylphenylhydroxylamine
  applications 102
  preparation 45
Separation of elements by solvent extraction with BPHA 158–192
Scandium
  amperometric determination 200
  gravimetric determination 60
  separation from lanthanides 162

Scandium (*cont.*)
  titrimetric determination 194
Stability constants of TBPHA and BPHA chelates 49
N-substituted
  arylhydroxylamines 47, 48
  hydroxylamines 37, 38
  phenylhydroxylamines 33

Thallium extraction, as a function of pH and BPHA concentration 180
Thiobenzoylphenylhydroxylamine (TBPHA)
  applications 103–105
  precipitation reactions 51, 52
  preparation 48
  reactions with metal ions 50
  spectrophotometric reagent 136
  stability constants of chelates 49
N-2-Thiophenecarbonyl-*p*-tolyhydroxylamine (TTHA)
  preparation 40
  spectral characteristics 41
  spectrophotometric reagent 136
Thorium
  determination by direct weighing 64
  extraction and separation from lanthanides 164
  gravimetric determination 63
Thorium and cerium, gravimetric determination 62
Thorium and uranium, separation from lanthanum 159
Tin, gravimetric determination 59
Tin and antimony, separation from indium 171
Titanium
  amperometric determination 198
  extraction and determination as titanium-BPHA-thiocyanate complex 128
  gravimetric determination 85, 102
  separation and determination by weighing the complex 87
  spectrophotometric determination 122–126, 141, 150, 153
Titanium and niobium, spectrophotometric determination 140
Titanium and vanadium, in uranium compounds, spectrophotometric microdetermination 125
Tungsten
  extraction 164
  gravimetric determination 78

Uranium(VI), gravimetric determination 72

Vanadium
  extractive separation 110
  spectrophotometric determination 109, 114, 136, 138, 142–144, 149, 154
  titrimetric determination 194
Vanadium, spectrophotometric determination when present in
  aluminium 116
  brine 151
  chrome-magnesite refractory 110
  chrome-vanadium steel 110
  chromite 110
  high speed steel 110
  ilmenite and rutile 119
  iron ores 120
  magnetite, ilmenite, chromite and igneous rocks 118
  plutonium-vanadium alloys 117
  rocks 116
  silicate rocks and minerals 148
  steel and petroleum products 115
  steels and ores 151
  titanium tetrachloride 110

Zinc, solvent extraction with BPHA 182
Zirconium
  amperometric determination 199
  extractive determination 165
  gravimetric determination 67, 68
  spectrophotometric determination 131
  titrimetric determination 198

OTHER TITLES IN THE SERIES IN ANALYTICAL CHEMISTRY

Vol. 1. WEISZ—Microanalysis by the Ring Oven Technique.
Vol. 2. CROUTHAMEL—Applied Gamma-ray Spectrometry.
Vol. 3. VICKERY—The Analytical Chemistry of the Rare Earths.
Vol. 4. HEADRIDGE—Photometric Titrations.
Vol. 5. BUSEV—The Analytical Chemistry of Indium.
Vol. 6. ELWELL and GIDLEY—Atomic Absorption Spectrophotometry.
Vol. 7. ERDEY—Gravimetric Analysis Parts I–III.
Vol. 8. CRITCHFIELD—Organic Functional Group Analysis.
Vol. 9. MOSES—Analytical Chemistry of the Actinide Elements.
Vol. 10. RYABCHIKOV and GOL'BRAIKH—The Analytical Chemistry of Thorium.
Vol. 11. CALI—Trace Analysis for Semiconductor Materials.
Vol. 12. ZUMAN—Organic Polarographic Analysis.
Vol. 13. RECHNITZ—Controlled-potential Analysis.
Vol. 14. MILNER—Analysis of Petroleum for Trace Elements.
Vol. 15. ALIMARIN and PETRIKOVA—Inorganic Ultramicroanalysis.
Vol. 16. MOSHIER—Analytical Chemistry of Niobium and Tantalum.
Vol. 17. JEFFERY and KIPPING—Gas Analysis by Gas Chromatography.
Vol. 18. NIELSEN—Kinetics of Precipitation.
Vol. 19. CALEY—Analysis of Ancient Metals.
Vol. 20. MOSES—Nuclear Techniques in Analytical Chemistry.
Vol. 21. PUNGOR—Oscillometry and Conductometry.
Vol. 22. J. ZYKA—Newer Redox Titrants.
Vol. 23. MOSHIER and SIEVERS—Gas Chromatography of Metal Chelates.
Vol. 24. BEAMISH—The Analytical Chemistry of the Noble Metals.
Vol. 25. YATSIMIRSKII—Kinetic Methods of Analysis.
Vol. 26. SZABADVÁRY—History of Analytical Chemistry.
Vol. 27. YOUNG—The Analytical Chemistry of Cobalt.
Vol. 28. LEWIS, OTT and SINE—The Analysis of Nickel.
Vol. 29. BRAUN and TÖLGYESSY—Radiometric Titrations.
Vol. 30. RUŽIČKA and STARÝ—Substoichiometry in Radiochemical Analysis.
Vol. 31. CROMPTON—The Analysis of Organoaluminium and Organozinc Compounds.
Vol. 32. SCHILT—Analytical Applications of 1,10 Phenanthroline and Related Compounds.
Vol. 33. BARK and BARK—Thermometric Titrimetry.
Vol. 34. GUILBAULT—Enzymatic Methods of Analysis.
Vol. 35. WAINERDI—Analytical Chemistry in Space.
Vol. 36. JEFFERY—Chemical Methods of Rock Analysis.
Vol. 37. WEISZ—Microanalysis by the Ring Oven Technique. (2nd Edition—large and revised.)
Vol. 38. RIEMAN and WALTON—Ion Exchange in Analytical Chemistry.
Vol. 39. GORSUCH—The Destruction of Organic Matter.

Vol. 40. MUKHERJI—Analytical Chemistry of Zirconium & Hafnium.
Vol. 41. ADAMS & DAMS—Applied Gamma Ray Spectrometry (Second edition).
Vol. 42. BECKEY—Field Ionization Mass Spectrometry.
Vol. 43. LEWIS and OTT—Analytical Chemistry of Nickel.
Vol. 44. SILVERMAN—Determination of Impurities in Nuclear Grade Sodium Metal.
Vol. 45. KUHNERT-BRANDSTATTER—Thermomicroscopy in the Analysis of Pharmaceuticals.
Vol. 46. CROMPTON—Chemical Analysis of Additives in Plastics.
Vol. 47. ELWELL & WOOD—Analytical Chemistry of Molybdenum and Tungsten.
Vol. 48. BEAMISH & VAN LOON—Recent Advances in the Analytical Chemistry of the Noble Metals.
Vol. 49. TOLGYESSY, BRAUN & KYRS—Isotope Dilution Analysis.

OHIO UNIVERSITY LIBRARY

Please return this book as soon as you have finished with it. In order to avoid a fine it must be returned by the latest date stamped below.

CF